ENCYCLOPAEDIA OF HORTICULTURE AND ALLIED TERMS

Volume -2

Foreword

Contrary to the popular belief as a subject of mere aesthetic gardening, Horticulture emerged to be a very promising multidisciplinary field with unbound potentialities assuring nutritional as well as economic security throughout the globe. With the advancement of science, its area and scope has been widened both in terms of concept and dimension of newer areas with newer terminologies to be understood for better comprehension. The present book *Encyclopaedia of Horticulture and Allied Terms* by Dr. A.B. Sharangi, Associate Professor, Bidhan Chandra Krishi Viswavidyalaya, is a welcome effort in this direction.

There are many revered books available with this theme in various names and forms, but very few books like the present one are there which cover almost the entire range of terms associated with Horticulture and allied sciences. I like to congratulate Dr. Sharangi for his maiden attempt in compiling such a valuable book.

I am confident that the book as such can meet the long-felt need of a well balanced encyclopaedia of this kind for a considerable period of time. I hope the book will be accepted widely by the students, teachers, researchers, amateur gardeners, policy makers, corporate personnel and all individual readers having keen interest on the diverse fields of Horticulture as theory as well as in practice.

June 5, 2008

Prof. (Dr.) M.G. Som
Former Vice Chancellor,
Bidhan Chandra Krishi Viswavidyalaya
(State Agriculture University),
West Bengal, India

Contents

Foreword *v*

Prologue *xi*

– Volume 1 –

Chapter 1
Horticultural Terms **1**

Chapter 2
Terms of Botany **29**

Chapter 3
Manures and Fertilizer Terms **63**

Chapter 4
Biotechnology and Genetic Engineering Terms **79**

Chapter 5
Pesticide Terms **289**

– Volume 2 –

Chapter 6
Plant Pathology Terms **305**

Chapter 7
Plant Breeding & Genetics Terms **353**

Chapter 8
Soil Science Terms **365**

Chapter 9
Pollen & Spore Terms 373

Chapter 10
Water and Water-Related Terms 413

Chapter 11
Quality Management Terms 429

Chapter 12
Crop Science Terms 447

Chapter 13
Rose Glossary 483

Chapter 14
Tea, Coffee and Wine Terms 493

Chapter 15
Statistical Terms 521

Chapter 16
Entomology Terms 531

Chapter 17
Instruments, Implements and Tools Used in Horticulture 541

Chapter 18
Bonsai and Flower Arrangement Terms 551

Chapter 19
Natural and Synthetic Chemicals Used in Horticulture 555

Chapter 20
Some Medicinal/Medical Terms 563

Chapter 21
Some Special Terminologies Collected from Different Sources 569

References 595

Prologue

Agriculture gave the primitive man a settled life on earth towards growth and development of civilization. The search for food with the creation of green life slowly made him satisfied, rational and social. Agriculture, by its nature, involves self-conscious attempts by humans to change or "manage" natural ecosystems. Horticulture initially started as aesthetic agriculture in a confined garden, slowly changing its stand towards high value high return crops per unit area, time and input with the ultimate aim for food and nutrition security of the ever-increasing population of the entire globe particularly in the developing countries. India is no exception to this pragmatic trend to assimilate and welcome horticulture in her vision and mission. Harnessing the compelling strength in biodiversity, India is on its way to becoming the global hub for fresh horticultural produce outsourcing with diverse agro climatic zones, and a wealth of native knowledge contributing to the sector's development. Our country also shares a warm relationship with several gross importers of horticultural produce.

Horticulture, itself is a unique subject which includes the broad and diverse areas of Botany, Plant Pathology, Entomology, Genetics, Breeding, Biotechnology, Genetic Engineering, Soil Science, Chemistry, Farm management, Economics, Sociology, Statistics, Medicine, Agricultural Engineering, Agrometeorology, etc in general and Pomology, Olericulture, Spices, Tea, Coffee, Rose, Wine, Post Harvest Technology, Processing, Bonsai, Flower arrangement, Pollen/Spore, Pesticides, Manures/Fertilizers, Natural/Synthetic chemicals, Instruments/Implements/Tools, Quality management, Water, etc in particular. The understanding of the entire area is more or less impossible without the understanding of the specific terminologies associated with each subject concerned. Few chapters have been placed seperately (*e.g.* Chapters 13, 14 and 21) with a due consideration of their relative importance and not placed in any common chapter like the others.

In spite of a good number of publications in horticultural terms, no concerted attempt has been made to satisfy the need of the interested readers to provide a comprehensive output of the entire areas related to horticulture. The present book "*Encyclopedia of Horticulture and Allied Terms*" is a humble attempt to bring all the important and related terms together. Each term has been defined with utmost care, accuracy and completeness to provide a rational understanding and which are readily accessible for the students, teachers, amateur gardeners, corporate sectors,and all professionals associated with Horticulture.

An encyclopedia can never be original in its strictest sense, but a collection of works and compilation of others. Many terms and their definitions have been derived from books, journals, websites collected from different sources. The author is grateful to them for rendering such indirect help and acknowledging each and everyone. My special thanks are due to Seema, my wife and Avijoy, my son for their silent help during the wee hours of manuscript preparation. I am realy indebted to Dr. P.K. Sahu, Reader, Agricultural Statistics, BCKV for kindly correcting the statistics section. There is every possibility of printing, typographical or other errors as the entire volume came into reality within a very short spell of time. Inspite of every effort to avoid duplications of terms in different chapters, there may be many such instances present in the book. Constructive criticism, valuable comments and all sorts of suggestions can only make the publication complete in its future appearances.

Dr. A.B. Sharangi

Chapter 6
Plant Pathology Terms

Abaxial–Directed away from the stem of a plant; pertaining to the lower surface of a leaf.

Abiotic–Non-living; of non-biological origin.

Abscission–Of plants: The shedding of leaves or other parts as the result of physical weakness in a specialized layer of cells (abscission layer) that develops at the base.

Acerose–Needle-shaped, like pine needles.

Acervulus–A mass of closely clustered conidiophores and conidia not covered by fungal tissue, initially subcuticular or subepldermal but eventually exposed. (pl. **acervuli**)

Acropetal–Describes the development of structures (such as spores) in succession from the base towards the apex.

Actinomycete–Any member of the order Actinomycetales; the name is often used to refer specifically to those species which form mycelium. (pl. **actinomycetes**).

Actinomycetes–(literally "ray fungi") filamentous bacteria that have sometimes been classified as Fungi Imperfecti. Actinomycetes typlcally are saprobes (especially in soil) but a few are pathogenic to man, animals, and plants.

Acute–1. Developlng suddenly, severe (with reference to disease symptoms). 2. Less than 90 degrees (with reference to an angle).

Acute symptoms–Search in shock symptoms

Adanal (Nematology)–Pertaining to a bursa that does not envelop the entire tail.

Adaxial–Directed toward the stem of a plant; pertaining to the upper surface of a leaf.

Adjuvant–Material added to improve some chemical or physical property (*e.g.*, of a plant protectant) or a biological property (*e.g.*, to improve antibody response to an antigen).

Adnate (Mycology)–Of gills or tubes broadly attached to the stipe; attached by nearly the entire width of the gills or layer of tubes.

Aecidiospore–Search in aeciospore.

Aecidium–An aecium with a cup-like outer wall. (pl. **aecidia**).

Aeciospore–A dikaryotic "transfer" spore of the Uredinales, formed in an aecium on the alternate host in macrocyclic rusts and infecting only the primary host; spores of Stage I in heteroecious or autoecious rusts.

Aecium–In the Uredinales the first sorus that is formed after plasmogamy and bears binucleate aeciospores (Stage I heteroecious or autoecious rusts). (pl. **aecia**.)

Aerobe–An organism that requires free oxygen for resplration.

Aerobic–With the qualities of an aerobe.

Agar–Mixture of polysaccharides derived from red algae that forms a gel at temperatures below about 40°C. Used as a support medium, when supplemented by appropriate buffers and/or nutrients and other ingredients, for the production of microbial cultures, overlaying tissue culture cells, electrophoresis, etc.

Agarose–One of the constituents of agar. Often used in preference to agar because it gels at a lower temperature and does not contain the inhibitors of virus growth frequently present in agar. It is also used widely in gel electrophoresis because it has a more uniform pore size than that of agar.

Agent of disease–An organism or abiotic factor that causes disease; a pathogen.

Agent of inoculation–That which transports inoculum from its source to or into the infection court (*e.g.*, wind, splashing rain, insects, humans).

Agglutination–The formation of insoluble aggregates following the combination of antibodies with cells or other particulate antigens or with soluble antigens bound to cells or other particles or following the combination of soluble (or particulate) antigens with cell-bound or particle-bound antibodies.

Aggressiveness–Of a plant pathogen: relative ability to colonize and cause damage to plants. Search in virulence. (Note: The Federation of British Plant Pathologists, now the British Society for Plant Pathology, has rejected this term and considered it to be synonymous with pathogenicity.)

Alae (Nematology)–Expansions or projections formed by a longitudinal thickening of the cuticle of a nematode. Cervical alae are confined to the anterior region of nematodes parasitic in animals. Caudal alae occur in the posterior region of males in a number of genera. Longitudinal alae, usually four, extend the length of the body sublaterally.

Allantoid–Of spores: sausage-shaped; somewhat curved, with rounded ends.

Allele. (allelomorph)–Any of one or more alternative forms of a given gene; both (or all) alleles of a given gene are concerned with the same trait or characteristic, but a particular allele codes for a product qualitatively and/or quantitatively different from that coded by other alleles of that gene.

Allelomorph–Search in allele.

Alternate host–One of two kinds of plants on which a parasitic fungus (*e.g.*, a rust) must develop to complete its life cycle.

Alternative host–A plant other than the main host that a parasite can colonize; alternative hosts are not required for completion of the developmental cycle of the parasite.

Amerospore–A non-septate spore.

Amerosporous–Having one-celled spores.

Amphid–In nematodes: A chemo-sensory organ, occurring laterally in pairs, located in the anterior region. Sometimes it is also called as lateral organs.

Amphidelphic–In nematodes: Having two ovaries, one directed anteriorly and the other posteriorly.

Amphigynous–In fungi: Having an antheridium through which the oogonial initial grows.

Amphimictic–In nematodes: 1. Reproduction in which sperm and eggs come from separate individuals (cross-fertilization). 2. Capable of interbreeding freely and of producing fertile offspring.

Amphiospore–Urediniospore with thickened walls and capable of hibernating.

Amphispore–Amphiospore.

Amplification–Production of multiple coples of a DNA sequence, either in vivo or in vitro, starting with one or a few coples.

Ampulliform–Flask-like in form.

Amyloid–Of spore walls, spore ornamentation, hyphal walls, ascus tips, etc.: Staining blue or grayish to blackish-violet in Melzer's reagent, presumably because of the presence of starch or a starch-like compound.

Anaerobic–Not requiring free, molecular oxygen for resplration.

Anamorph–An asexual (usually conidial) stage in the life cycle of a fungus.

Anastomosis (= hyphal fusion)–Fusion of somatic hyphae; characteristic of most Dikaryomycota. (Pl. **anastomoses.**)

Aneuploid–Chromosome constitution differing from the usual diploid constitution by loss or duplication of chromosomes or chromosomal segments.

Angiocarpous–Of basidium-producing organs: Hymenial surface at first exposed but later covered by an incurving plleus margin and/or excresences from the stipe.

Annule–Thickened interval between transverse striae in the cuticle of a nematode.

Annulus–A membraneous skirt surrounding the stipe of a hymenomycete or gasteromycete. (pl. **annuli**)

Antagonism–An ecological association between organisms in which one or more of the participants is harmed or has its activities limited.

Antagonist–An agent or substance that counteracts the action of another.

Antagonistic symbiosis–Parasitism; one organism of an association beefits at the expense of the other.

Antheridium–A male gametangium. (pl. **antheridia.**)

Anthocyanescence–Having reddish-purple color in tissues that are normally green; often a symptom of plant disease appearing in the nearly dead margins around completely dead spots in green leaves.

Anthracnose–Any of various plant diseases, particularly those caused by fungi of the Melanconiales, in which discrete, dark-colored, necrotic lesions develop on the leaves, stems, and/or fruits.

Antibiosis–An association between two organisms that is detrimental to the vital activities of one of them.

Antibiotic–Substance used to inhibit the growth of micro-organisms, including bacteria and fungi.

Antibody–Any immunoglobulin molecule produced in direct response to an antigen and which can combine specifically, non-covalently, and reversibly with the antigen which elicited its formation.

Antigen–Molecule of carbohydrate or protein which stimulates the production of an antibody, with which it reacts specifically.

Antiserum–The serum from a vertebrate that has been exposed to an antigen and which contains antibodies that react specifically with the antigen. (pl. **antisera.**)

Apiculate–Having an apical point or aplculus.

Apiculus–A short, sharp, but not stiff, point, usually as the bud of a spore. (pl. **apiculi.**)

Apothecium–An ascus-bearing structure (ascocarp) in which the ascus-producing layer (hymenium) is not covered by fungal tissue at maturity. (pl. **apothecia.**)

Appressorium–An enlargement on a hypha or germ tube that attaches itself to the host before penetration takes place. (pl. **appressoria.**)

Arbuscule–Of vesicular-arbuscular mycorrhizae: a much-branched, microscoplc haustorial structure of the fungal symbiont that forms within living cortlcal cells of the root. The interface of the arbuscule with the plant protoplast is a site of exchange of nutrients and growth-regulating chemicals.

Arcuate–Curved like a bow.

Areolated–Divided into small spaces or areolations; usually pertains to the cuticle of a nematode.

Areolation–A condition in which the transverse body annulation of a nematode traverses the lateral field.

Arthrospore–A spore resulting from the fragmentation of a hypha, as in the conidial stage of many Basidio-mycetes.

Ascigerous–Of fungi: having asci.

Ascocarp–Ascospore-bearing, multi-cellular sporocarp formed by a member of the Ascomycotina.

Ascoconidium–A budded spore arising from an ascospore within an ascus.

Ascogenous hypha–The restricted dikaryophase of many Ascomycetes; a dikaryotic hypha that grows out from the fertilized ascogonium.

Ascogonium–In Ascomycetes: the female gametangium; it may be unicellular or multicellular, simple or complex in form. (pl. **ascogonia.**)

Ascohymenial–Of, pertaining to, or having the characteristics of the Ascohymeniales.

Ascohymeniales–Ascomycetes having asci (and paraphyses) developlng as a hymenium and not in a preformed stroma.

Ascolocular–Of, pertaining to, or having the characteristics of the Loculoascomycetes, the loculate Ascomycetes.

Ascoma–Synonymous with ascocarp. (pl. **ascomata.**)

Ascomycete–Fungus of the subdivision Ascomycotina.. (In some taxonomic schemes these fungi form the class **Ascomycetes.**)

Ascomycotina–A subdivision of fungi characterized by the formation of sexually derived spores (ascospores) in asci. (**Ascomycetes, Ascomycota**)

Ascophore–A structure bearing asci (*e.g.*, an ascus-producing hypha). **ascospore**. A spore borne in an ascus.

Ascostroma–A fructification of the Ascomycetes consisting of an undifferentiated mass of tissue or stroma on or in which the asci are developed. (Pl. **ascostromata.**)

Ascus–A cell that is the site of meiosis and in which endogenous spores (usually meiospores but sometimes also ascoconidia) are formed. (pll. **asci.**)

Aseptate–Lacking septa (nonseptate).

Asporogenous–Not capable of forming spores.

Autoecious–Capable of completing a life cycle on one host.

Autotrophic–Capable of growth independent of outside sources of nutrients or growth factors.

Auxotroph–A strain of microorganism lacking the ability to synthesize one or more essential growth factors.

Avirulent–Not exhibiting virulence; nonpathogenic.

Avoidance–Principle of plant disease control marked by deliberate actions to take advantage of environmental factors and time unfavorable for disease development.

Axial–Belonging to, around, in the direction of, or along an axis.

Azygospore–A parthenogenetic zygospore; formed by some vesicular-arbuscular mycorrhizal fungi (family Endogonaceae).

Bacillar–Shaped like a short rod.

Bacilliform–Rod-shaped.

Bacillus–1. A member of the genus *Bacillus*. 2. Any rod-shaped bacterial cell, *i.e.*, a cell whose length is two or more times greater than its width.

Backcross–To cross (mate) with one parent.

Bactericidal–Lethal to bacteria.

Bactericide–A substance that kills bacteria.

Bacteriocin–A protein antibiotic, one or more types of which can be produced and "exported" (excreted) by certain strains of bacteria.

Bacteriophage–A virus that replicates inside a bacterium.

Bacteriostati–Able to inhibit the growth and reproduction of at least some types of bacteria.

Ballistospore–A spore that when mature is actively projected.

Basidiocarp–A sporocarp produced by a member of the Basidiomycotina and which bears basidiospores.

Basidiole–A structure in the hymenium of a member of the Basidiomycotina that is morphologically similar to a basidium without sterigmata. It may be an immature basidium or a permanently sterile structure in the hymenium.

Basidioma–Search in basidiocarp. (pl. **basidiomata.**)

Basidiomycete–A fungus of the Basidiomycotina.

Basidiomycetes-Search in Basidiomycotina.

Basidiomycota–Search in Basidiomycotina.

Basidiomycotina–A subdivision of fungi characterized by the formation of basidiospores on basidia.,

Basidiospore–An exogenous sexual spore (meiospore) borne on a basidium.

Basidium–A cell in which karyogamy and meiosis take place and which bears exogenous spores of sexual origin. (pl. **basidia**.)

Basipetal–Successive from apex to base.

Bifurcate–Dividing into two branches.

Binary fission–Fission in which two cells, usually of similar size and shape, are formed by the growth and division of one cell.

Bioassay–Any quantitative procedure in which a given organism is used for assay purposes.

Biocontrol–Search in biological control.

Biological control–The deliberate use by humans of one species of organism to eliminate or control another.

Biotroph–An organism that derives nutrients from the living tissues of another organism (its host).

Biotype–A subspecies of organism morphologically similar to but physiologically different from other members of the species.

Bipolar–At both ends or poles.

Biseptate–Having two septa.

Bitunicate–Having two walls.

Blastic–One of two basic kinds of conidiogenesis; there is a marked enlargement of a recognizable conidium before it is delimited by a septum. (Search in thallic.)

Blasting–A symptom of plant disease characterized by shedding of unopened buds; classically, the failure to produce fruit or search ind.

Blastoconidium–Search in blastospore.

Blastospore–A spore that arises by budding, as in yeasts.

Blight–A disease characterized by rapld and extensive death of plant foliage. A general term applied to any of a wide range of unrelated plant diseases. (*e.g.*, chestnut blight, fireblight, late blight, halo blight)

Blot–Search in blotting.

Blotch–A disease characterized by large, and irregular in shape, spots or blots on leaves, shoots, and stems.

Blotting–Following electrophoresis: the transfer of nucleic acids and/or proteins from a gel strip to a specialized, chemically reactive matrix on which the nucleic acids, etc. may become covalently bound in a pattern similar to that present in the original gel.

Broadcast application-Application by spreading or scattering on the soil surface.

Broom-In plant pathology: A symptom in which lateral branches proliferate in a dense cluster on the main branch (witches'-broom).

Broth–In microbiology: Any of a variety of liquid media, especially nutrient broth or any liquid medium based on nutrient broth and/or hydrolysed protein.

Buccal capsule–In nematodes: Structure connecting the oral opening with the anterior portion of the esophagus. The buccal capsule (also called the stoma) is subject to great variation among different nematodes.

Bullae–In nematodes: Blisterlike prominences near the vulval fenestra of some Heteroderidae.

Bursa–In nematodes: Caudal alae of males used to clasp the female during copulation.

Caeoma–An aecium in the Uredinales that is not surrounded by a peridium; from the form genus *Caeoma*. (pl. **caeomata**.)

Callus–A mass of thin-walled, undifferentiated plant cells, developed as the result of wounding or culture on nutrient media.

Canker–An imprecise term usually used for a plant disease characterized (in woody plants) by the death of cambium tissue and resulting loss and/or malformation of bark, or (in non-woody plants) by the formation of sharply delineated, dry, necrotic, localized lesions on the stem. The term "canker" may also be used to refer to the lesion itself, particularly in woody plants.

Capillitium–A mass of sterile fibers interspersed among spores within a sporocarp (in the Gasteromycetes and Myxomycota). (pl. **capillitia**.)

Capitulum–In nematodes: Medial ventral sclerotization of the spicular pouch.

Capsid–The protein shell that surrounds the virus nucleic acid.

Capsule–In bacteria: A layer of material external to but contiguous with the cell wall.

Cardia–In nematodes: Valvular apparatus connecting the esophagus and intestine. Sometimes called the cardiac valve or esophago-intestinal valve.

Carlavirus–(Siglum of carnation latent virus.) Member of a group of plant viruses with slightly flexuous, rod-shaped particles containing a single molecule of linear RNA, most of which are transmitted by aphids in a noncirculative manner.

Carmovirus–(Siglum of carnation mottle virus.) Member of a group of plant viruses with small, isometric particles containing a single molecule of linear RNA, transmitted in nature through soil and (rarely) by an insect vector.

Carrier–An organism that bears an infectious agent internally but shows no marked symptoms of the disease caused by that agent.

Caudal–In nematodes: Pertaining to or located near the posterior region or tail.

Causal agent of diseas–That which is capable of causing disease.

Cell cycle–The period from one cell division to the next.

Cephalic–In nematodes: Pertaining to or located near the head.

Cephalids–In nematodes: Two structures (posterior and anterior) situated in the cephalic region and extending in a complete circle around the body; possibly part of the nervous system. Sometimes called hypodermal commisures.

Cfu–Colony-forming unit.

Chemotherapy–The use of chemical(s) (*e.g.*, antibiotics or fungicides) for the treatment of a disease.

Chlamydospore–A thick-walled, non-sexual spore; a transformed hyphal cell.

Chloranemia–The necrotic symptom of yellowing; a loss of chlorophyll.

Chlorosis–The loss of chlorophyll from the tissues of a plant, resulting from microbial infection, viral infection, the action of certain phytotoxins, the lack of light, to magnesium or iron deficiency, etc. Chlorotic tissues commonly appear yellowish.

Chlorotic–Search in chlorosis.

Chord–In nematodes: A longitudinal internal thickening of the hypodermis.

Circulative transmission–Virus transmission characterized by a long period of acquisition of the virus by a vector, a latent period of several hours before the vector is able to transmit the virus, and retention of the virus by the vector for a long period, usually several days. (Also termed **persistent transmission**)

Cirrhus–(Also **cirrus.**) 1. A mass of spores in the form of a ribbon or tendril, forced from the fruiting body of a fungus. 2. A discrete group of somatic cilia (several to over 100) that act primarily as a unified locomotive organelle; the typlcal cirrus is conical. (pl. **cirrhi** or **cirri.**)

Clamp connection–A recurving outgrowth of a cell that, at cell division, acts as a bridge to allow passage of one of the products of nuclear division into the penultimate cell, thereby assuring maintenance of the dikaryotic condition (of members of the Basidiomycotina).

Clavate–Club-shaped.

Cleistothecium–An ascocarp with the asci surrounded by fungal tissue and without regularly formed openings. (pl. **cleistothecia.**)

Cloaca–In nematodes: A common duct or cavity in which the digestive and reproductive systems terminate in males.

Clone–1. (n.) (a) A population of recombinant DNA molecules all carrying the same inserted sequence; (b) a colony of micro-organisms containing a specific DNA fragment inserted into a vector; (c) a population of cells or organisms of identical genotype.

2. (v.) (a) the use of in vitro recombination techniques to insert a particular DNA sequence into a vector; (b) the selection of a unique virus isolate from individual plaques, pocks or lesions or by limiting dilution; (c) the vegetative propagation of an organism to produce a population of identical individuals.

Cloning–An *in vitro* procedure in which a particular sequence of DNA (*e.g.*, a gene) is reproduced in large amounts by inserting ("splicing") it into a suitable replicon, introducing the resultant recombinant (hybrid) molecule into a cell in which it can replicate, and finally growing the cells in culture.

Closterovirus–(from Greek *kloster*, "thread") Member of a group of plant viruses with very long, flexuous, rod-shaped particles containing a single molecule of linear RNA, some members of which are transmitted by whiteflies.

Cluster cup–Aecidium.

Coalesce–To merge or grow together into a similar but larger structure.

Coccus–A spherical (or near-spherical) bacterial cell. (pl. **cocci**).

Codon–A particular sequence of three nucleotides in mRNA coding for an amino acid.

Coelomycetes–A group of the Deuteromycetes producing pycnidia or acervuli.

Coenocyte–A multinucleate cell; a protoplast in which the nuclear divisions have not been followed by cytoplasmic cleavage.

Coenocytic–Multinucleate or without cross walls. Search in syncytium.

Coenozygote–A cell containing more than one zygote.

Commensalism–Symbiosis in which neither organism is injured; one or neither may be benefited.

Commisure–In nematodes: Connecting bands of nerve tissue.

Comovirus–(Siglum of cowpea mosaic virus). Member of a group of multicomponent plant viruses with small, isometric particles containing two linear RNA species, readily transmitted mechanically and by beetles.

Compartmentalization–In trees: the processes that result in isolation of wounded or diseased xylem from normal xylem by the formation of chemically and anatomically specialized tissue around the damaged zone.

Competition–A more or less active demand on the part of two organisms for some commodity (space, food, etc.) that is inadequate to provide for all organisms present.

Conidiogenesis–Conidium formation.

Conidiogenous cell–A conidium-producing cell.

Conidioma–A specialized, multi-hyphal structure bearing conidia. (pl. **conidiomata.**)

Conidiophore–A hypha, often specialized in structure, that bears one or more conidia.

Conidium–A thin-walled, asexual spore that is borne exogenously on a conidiophore and is deciduous at maturity. (Search in endoconidium.) (pl. **conidia.**)

Conjugate–To carry out conjugation.

Conjugation–In general, any of various sexual processes in microorganisms in which gene transfer follows the establishment of direct contact between two (or more) cells which typlcally show little or no morphological differentiation from vegetative cells. In bacterial conjugation, one bacterium (the "male" or donor cell) transfers DNA to another (the "female" or reciplent cell) while the cells are in physical contact; a reciplent that has received DNA from a donor is called a transconjugant.

Conk–The basidiocarp of a wood-decaying fungus, usually a polypore.

Context–The inner or body tissue of a fruit body which supports the hymenophore in the larger and especially the plleate species of Hymenomycetes.

Control–Economic reduction of crop losses caused by plant diseases.

Coremium–Search in synnema. (pl. **coremia**.)

Cornute–Horned; horn-like.

Corpus–In nematodes: The anterior cylindrical part of the esophagus. The basal region of the corpus at times may be swollen to form a bulb.

Coryneform–1. Essentially rod-shaped with one end thickened or bulbous. 2. A name applied, loosely, to any Gram-positive, asporogenous, pleomorphic rod-shaped bacterium; as such it covers bacteria from a range of genera.

Cosmid–A plasmid into which has been inserted the *cos* site of bacteriophage.

Crop rotation–The practice of growing a sequence of different crops on the same land in successive years or seasons; done to replenish the soil, curb pests, etc.

Cross-protection. The protection conferred on a host by infection with one strain of a virus that prevents infection by a closely-related strain.

Crozier–A recurved hook at the tip of an ascogenous hypha, the penultimate cell of which will become the ascus.

Crozier formation–Process of ascus development from coiled tips of ascigerous hyphae.

Crustaformeria–In nematodes: Glandular region of the distal part of uterus that may play a role in the formation of the egg envelope; sometimes called the quadricolumella.

Cucumovirus–(Siglum of cucumber mosaic virus). Member of a group of multicomponent plant viruses with isometric (icosahedral) particles consisting of three linear RNA species (RNAs 1, 2, and 3), transmitted by sap and in nature by aphids in a noncirculative manner.

Cultivar–A cultivated plant variety or cultural selection.

Culture–1. To grow an organism. 2. the resulting growth. Usually on artificial medium.

Culture collection–A repository of cultures of characterized viruses, bacteria, and other organisms. Used for reference and comparison with new isolates.

Cupulate–Cup-shaped.

Cuticle–1. A thin, waxy layer on the outer wall of epldermal cells consisting primarily of wax and cutin.

2. Noncellular exterior covering of nematodes.

Cutin–An insoluble polymer that, embedded in waxes, forms the cuticle covering the epldermal cell walls in the aerial parts of higher plants.

Cutinolytic–Of certain enzymes: able to digest cutin.

Cutis–Of basidiocarps of certain wood-decaying fungi: the outer layer consisting of compressed hyphae parallel to the surface, sometimes with varnishlike incrustation.

cv–Cultivar.

Cylindrical–Of the stipe, spores, etc.: Of the same diameter throughout the length.

Cyst–In fungi: An encysted zoospore. In nematodes: the egg-containing carcass of dead adult females of the genus Heterodera or Globodera.

Cystidium–A sterile cell occurring among basidia and often projecting beyond the hymenium, differing morphologically from the basidium. (pll. **cystidia.**)

Cytokidamping-off–A disease that results in the collapse and death of search indlings before or after they emerge from the soil (pre-emergence and post emergence damping-off, respectively). Common causal agents include species of *Pythium* and *Rhizoctonia*.

Decay–The gradual decomposition of dead organic matter.

DEGO (Dorsal Esophageal Gland Outlet)–In nematodes: The point at which the dorsal gland empties into the lumen of the esophagus.

Deirid–In nematodes: Paired, porelike organs located in the lateral fields; in the vicinity of the nerve ring of many of the Tylenchoidea; believed by some workers to be sensory in nature.

Denticle–In nematodes: Minute tooth or "prickle". (pl. **denticles.**)

Dermatophyte–A parasitic fungus that attacks and causes a disease of the skin.

Desiccate–To dry out.

Determinate–Having a fixed, definite limit.

Deuteromycete–A member of the Deuteromycotina.

Deuteromycetes–Search in Deuteromycotina.

Deuteromycotina (Also **Fungi Imperfecti**)–A non-phylogenetic category originally created for fungi with no known sexual stage; the category still includes fungi with no known sexual stage, but it also includes the asexual stages of various fungi now known to have a sexual stage in the Ascomycotina or Basidiomycotina.

Dextrinoid-(Of spores, etc.) stained yellowish–or reddish-brown by Melzer's iodine.

Diagnostic–It is a distinguishing characteristic which is important in the identification of a disease or other disorders.

Dichotomous–Dividing into two equal branches.

Dictyospore–A spore divided by intersecting septa in more than one plane.

Dictyosporous–Having spores with cross and longitudinal walls.

Didelphic–Of nematodes: Possessing two complete genital tubes or ovaries.

Didymospore–A spore with one transverse septum.

Didymosporous–Having two-celled spores.

Dieback–Progressive death of shoots, branches and roots generally starting at the tip.

Differential host–A plant host that on the basis of disease symptoms serves to distinguish between various strains or races of a given plant pathogen.

Dikaryon–A pair of nuclei that associate and divide simultaneously.

Dikaryotic–The condition of containing a dikaryon.

Dimorphic–Producing two morphologically different forms.

Dimorphism–Existence of two morphologically different forms in one organism.

Dioecious–Having male and female reproductive structures on separate thalli.

Diorchic–In nematodes: Possessing two testes.

Diploid–Having a ploidy of two. Represented by 2n.

Disc–Of Valsa and related fungi: a more or less flat aplcal part of a stroma that protrudes above the bark surface; also, of Discomycetes: the exposed fertile portion of an apothecium.

Discomycete–A member of the Discomycetes.

Discomycetes–A group of the Ascomycetes in which the hymenium is exposed at maturity; one in which the fruiting body is an apothecium or discocarp.

Disease–An abnormal condition of a plant in which its physiology, morphology, and/or development is altered under the continuous influence of a pathogen.

Disease cycle–Of a disease caused by a biotic agent: the cyclical sequence of host and parasite development and interaction that result in disease, in reproduction or replication of the pathogen, and in the readying of a new generation of the parasite for infection.

Disinfectant–A physical or chemical agent that frees a plant, organ, or tissue from infection is said to be as disinfectant.

Disinfest–To kill pathogens that have not yet initiated disease, but that occur in or on such inanimate objects as soil, tools, etc., or that occur on the surface of such plant parts as search ind.

Disinfestant–An agent that kills or inactivates pathogens in the environment or on the surface of a plant or plant organ before infection takes place.

Dispersal–Spread of a pathogen within an area of its graphical range.

Dissemination–Search in dispersal.

DNA–Deoxyribonucleic acid.

DNAase-An enzyme that attacks bonds in DNA.

DNA fingerprinting–A laboratory technique in which the banding patterns of DNA fragments from two different individuals are compared.

DNA polymerase–An enzyme that synthesizes a daughter strand(s) of DNA (under direction from a DNA template). May be involved in repair or replication.

DNA replicase–A DNA-synthesizing enzyme required specifically for replication.

Dorsal–Back or upper surface.

Downy mildew–Plant disease caused by certain members of the Peronosporales. Downy mildews are characterized by the formation of superficial hyphal growth in which, typlcally, individual spore-bearing structures can be distinguished.

Echinate–Having sharply pointed spines.

Echinulate–Covered with small spines.

Ectoparasite–A parasite that remains external to the host's cells or tissues.

Ectosymbiosis–Symbiosis in which one member (microsymbiote) develops on the outside of the other member.

Ectotrophic–Refers to a mycorrhiza in which the mycelium forms an external covering on the root.

Edema–Search in oedema.

Eelworm–Nematode.

Effective dissemination–Synonymous with inoculation.

Effuse–Spreading out loosely or flat.

Electrophoresis–A procedure by means of which the members of a heterogenous population of charged particles can be separated by virtue of their dissimilar migration characteristics in an electric field.

Electroporation–A method by which nucleic acids or virus particles can be introduced into protoplasts or cells by creating transient pores in the plasma membrane using an electric pulse.

Elicitor–A molecule produced by the host (or pathogen) that induces a response by the pathogen (or host).

ELISA–Enzyme-linked immunosorbent assay. A highly sensitive immunoassay for specific antibodies or antigens.

Ellipsoid–Having every plane section an ellipse or a circle.

Ellipsoidal–Search in ellipsoid.

Emarginate–Of gills, notched near the stipe.

Enation–A symptom caused by certain plant viruses in which there are small outgrowths on the plant.

Encysted–Surrounded by a hard shell (cyst).

Endemic–Of a disease: Native to a particular place.

Endoconidiophore–A conidiophore that produces conidia within itself.

Endoconidium–A conidium produced endogenously in a hypha or conidiophore. (Pl. **endoconidia**.)

Endogenous–Arising from within the generating structure.

Endonuclease–A nuclease which cleaves phosphodiester bonds *within* a nucleic acid strand.

Endoparasite–A parasite that lives intracellularly or within the tissues of the host.

Endosymbiosis–Symbiosis in which one member (microsymbiote) lives within the other.

Endotrophic–Refers to a mycorrhiza in which the mycelium grows within the cortical cells of the root (*e.g.*, in orchids).

Enphytotic–A plant disease that causes about the same amount of damage each year.

Enzyme–A protein produced by living cells that can catalyze a specific organic reaction.

Epidemic–A change in the amount of disease in a population in time and space.

Epidemiology–1. The study of the interrelationships between a given pathogen, the environment, and groups or populations of the relevant hosts.

2. The study of epidemics.

Epidermis–The superficial layer of cells occurring on all plant parts.

Epinasty. Downward curling of a leaf blade resulting from more rapld cell growth on the upper side of a petiole than on the lower side; often a hyperplastic symptom of plant disease.

Epiparasite–An organism parasitic on another that parasitizes a third.

Epiphyte–An epiphytic organism.

Epiphytic–Growing externally on a plant without parasitizing it.

Epiphytotic–Search in epidemic.

Epiptygma–In nematodes: A vulval flap.

Epitype–A specimen selected as a standard for a species or lower taxon when all original material except for illustrations has been destroyed.

Eradicant–Any chemical agent that eliminates particular pathogen(s) from diseased plants treated with that agent.

Eradication–Control of plant disease by eliminating the pathogen after it is established or by eliminating the plants that carry the pathogen.

Ergot–Disease of certain grasses and cereals, especially rye, caused by *Claviceps purpurea*; also the spur-shaped sclerotium of *C. purpurea* that replaces the grain in a diseased inflorescence.

Erumpent–Breaking through the surface; bursting forth.

Escape–Failure of inherently susceptible plants to become diseased, even though disease is prevalent.

Esophagus–In nematodes: The portion of the alimentary canal between the buccal capsule, or stoma, and the anterior portion of the intestine.

Ethidium bromide–(2,7-diamino-10-ethyl-9-phenylphenanthridinium bromide). A trypanocidal, bacteriostatic dye that binds to DNA and fluoresces under near-ultraviolet light; used for tracking nucleic acids.

Etiolation–A phenomenon exhibited by plants grown in the dark: etiolated plants are pale yellow and have long internodes and small leaves.

Etiology–The study of cause; that phase of plant pathology dealing with the causal agent and its relations with the susceptible plant.

Exclusion–The principle of plant disease prevention in which the pathogen is prevented from entering a given region.

Excretory–In nematodes: A tube or canal, lined with cuticle, that leads to the excretory pore.

Excretory pore–In nematodes: The exterior opening of the excretory system, generally located on the ventral side of the body near the basal region of the esophagus; also known as the orifice of the cervical gland.

Exogenous–Arising on the outside of the generating structure.

Exonuclease–A nuclease that sequentially removes nucleotides from one end of a strand of nucleic acid.

Exopathogen–Nonparasitic organism whose extracellular toxic metabolites cause disease in plants.

Extracellular–Outside the cells.

Exudate–Material that has passed from within a plant structure to the outer surface or into the surrounding medium; as in leaf exudate, root exudate, etc.

Facultative anaerobe–Refers to an organism that normally grows aerobically but can grow anaerobically.

Facultative parasite–An organism that normally lives as a saprophyte but under certain conditions can live as a parasite.

Facultative saprophyte–A mainly parasitic organism with the ability to survive for a part of its life cycle as a saprophyte and be cultured on artificial media.

Falcate–(Of spores) sickle-shaped.

Fallow–Previously cultivated land kept free from crops or weeds during at least one growing season.

Fasciation–Hyperplastic symptoms characterized by a fusing (and flattening) of such plant organs as stems.

Fasciculation–Hyperplastic symptom characterized by a clustering of such plant organs as shoots into such structures as witches' brooms.

Fenestra–In nematodes: A window or transparent spot; in some Heteroderidae, a thin-walled region of the vulval cone.

Fenestrate–Having a fenestra.

Fenestration–The area in which the fenestra occurs.

Fermentation–Oxidation of certain organic substances in the absence of molecular oxygen.

Filiform–Thread-shaped.

Fission–Cell division by cleavage (splitting) of the cell into two parts.

Fitness–The ability of an organism to survive and reproduce; the ability of an organism to pass its genes to the next generation.

Flagellate–Having one or more flagella.

Flagellum–A whip-like appendage responsible for motility in the majority of motile bacteria and other protists, fungi, algae, etc. (pl. **flagella**.)

Flexure–A turn or fold.

Form genus–A non-phylogenetic category, equivalent to genus, distinguished on the basis of one or more morphological features. In the Deuteromycotina, form genera are used to classify anamorphs; such form genera are based primarily on the characteristics (including mode of development) of the conidia, conidiophores, and conidiomata.

Form species–A non-phylogenetic category, equivalent to species, distinguished on the basis of one or more morphological features of an anamorph, treated as if it were an independent entity, especially for indexing or identification purposes; of importance chiefly in the Ascomycetes and Uredinales.

Forma-specialis–Literally "special form". An infraspecific taxonomic rank in which the taxa are distinguished on a physiological basis, particularly on the basis of adaptation to (or pathogenicity for) one or more specific hosts. In mycology, *forma specialis* is a taxonomic rank lower than form, subvariety, variety and subspecies, and higher than physiological race. Abbreviated f. sp. (pl. f. spp.) (pl. **formae speciales**.)

Freeze-drying–Search in lyophilization.

Fructification–1. Synonymous with fruiting body. 2. The formation or development of a fruiting body.

Fruit body–Search in **fruiting body**.

Fruiting body–Any multi-hyphal structure that bears or contains spores.

Fumigant–A gas or volatile substance that is used to disinfest certain areas of various pests.

Fungicidal–Kills fungi.

Fungi imperfecti–Search in Deuteromycotina.

Fungicide–A chemical or physical agent that kills or inhibits the growth of fungi. (Note some substances termed "fungicides" are fungistatic in their action.

Fungistasis–Inhibition of fungal growth or reproduction that is not lethal.

Fungistat–A substance that prevents fungal growth without killing the fungus.

Fungistatic–Able to inhibit the growth and/or reproduction of at least some types of fungi.

Fungus–A eukaryotic, heterotrophic organism whose usually walled, threadlike cells absorb nutrients. (pl. **fungi**.)

Furcate–Forked.

Fusiform–Spindle-shaped; tapering at each end.

Gall–An abnormal plant structure formed in response to parasitic attack by certain microorganisms (bacteria, fungi, viruses) or insects. Galls may develop either by localized cell proliferation or increase in cell size.

Gametangium–A structure that contains gametes.

Gamete–A haploid cell or nucleus involved in sexual reproduction, during which two gametes fuse to form a zygote.

Gametogenesis–The development of gametes.

Gasteromycete–A member of the Gasteromycetes.

Gasteromycetes–The group of Basidiomycetes with spores borne in cavities within the fruit body.

Gel–A matrix of polyacrylamide, agarose or similar material in which the electrophoresis of molecules is carried out.

Gel electrophoresis–A type of electrophoresis in which the molecules in a sample moves through a gel composed of agarose or polyacrylamide.

Geminivirus–(From Latin *gemini*, "twins", for the typical double particles). Member of the only group of plant viruses with double particles, each particle containing circular or linear single-stranded DNA; this group is divided into two subgroups I and II (transmitted in a circulative manner by leafhoppers) and subgroup III (transmitted in a circulative manner by whiteflies or, in the case of some viruses in this group, mechanically transmitted).

Gene expression–The transcription of mRNA from the DNA sequence of a gene and the subsequent translation of that mRNA to give the protein gene product. Less strictly it can mean the transcription step alone.

Genetic code–The nucleotide sequence of a DNA molecule (or, in certain viruses, of an RNA molecule) in which information for the synthesis of proteins is contained.

Genital paplllae–In nematodes: Tactile or sensory organs located on the male tail.

Genital primordium–In nematodes: The initial cells of the reproductive system.

Genome–The genetic information for an organism, consisting (in the case of viruses) of one or more species of either RNA or DNA, but not both.

Genotype–The genetic constitution of an organism.

Genus–A taxonomic rank above species and below family; the generic name of an organism is the first of the binomial. (pl. **genera**.)

Germ tube–A short, hypha-like structure that develops from certain types of spores upon germination.

Germination–In bacteria or fungi, the process by which a spore gives rise to a vegetative cell or hypha.

Germination by repetition–On the germination of a spore, the formation of a secondary spore rather than a germ tube.

Gill–A hymenium-covered, plate-like appendage that hangs from the under surface of the basidiocarp of some Hymenomycetes.

Glaucous–Having a bluish gray waxy surface.

Globose–Spherical, or nearly so.

Gram reaction–The result of the Gram stain.

Gram stain–An important bacteriological staining procedure discovered empirically in 1884 by the Danish scientist Christian Gram. When bacteria are stained with certain basic dyes, the cells of some species (Gram-negative species) can be easily decolorized with organic solvents such as ethanol or acetone. Cells of Gram-positive species resist decolorization.

Gubernaculum–In nematodes: Spicule guide; sclerotized accessory plece.

Guiding ring–In nematodes: Sleeve-like structure that surrounds and guides the stylet in genera of the Dorylaimoidea. Position varies among the genera from near apex to posterior portion of stylet.

Gummosis–A plant disease in which the lesions exude a sticky liquid.

Guttation–Exudation of water from plants, particularly along the leaf margin.

Habitat–A place with a particular kind of environment suitable for the growth of an organism.

Haploid–Having a ploidy of one or a single set of chromosome.

Haustorium–A specialized branch of a hypha formed inside a host cell by certain plant-parasitic fungi (especially obligate parasites) in order to obtain nutrients. (pl. **haustoria**.)

Head–In nematodes: That portion anterior to the base of the stoma or stylet.

Helicospore–A cylindric, splral or convolute spore, usually septate.

Helicosporous–Having spiral or at least strongly curved, often septate, spores.

Hemizonid–In nematodes: Lens-like structure situated between the cuticle and hypodermal layer on the ventral side of the body just anterior to the excretory pore; generally believed to be associated with the nervous system.

Hemizonion–In nematodes: A companion structure to the hemizonid; it is smaller and located posterior to the hemizonid.

Heteroecious–Requiring more than one host species to complete a life cycle (*e.g.*, of Uredinales)

Heterogamy–Plasmogamy between morphologically different gametes.

Heterokaryon–A cell that contains genetically different nuclei or a thallus made up of such cells.

Heterokaryosis–The result of forming a heterokaryon of a fungus; the condition of a hypha or cell having two or more genetically distinct haploid nuclei.

Heterokaryotic–The condition of being a heterokaryon.

Heteroploid–Having a complement of chromosomes differing from that characteristic of the species.

Heterothallic–The condition of being self-sterile, requiring a partner for sexual reproduction.

Heterothallism–The phenomenon in which sexual reproduction requires the involvement of two different thalli.

Heterotopy–Hyperplastic symptom in which an organ develops in a position other than its normal one.

Heterotroph–An organism that obtains its food from other organisms, living or dead.

Heterotrophic–Requiring organic substrates for growth and development; being incapable of synthesizing required organic materials from inorganic sources.

Histopathology–The study of pathology of cells and tissues; the microscoplc changes characteristic of disease.

Holobasidium–A single-celled basidium. (pl. **holobasidia**.)

Holoblastic–When both outer and inner walls of the conidiogenous cell contribute to the formation of the blastoconidium.

Holocarpic reproduction–In fungi, reproduction in which the entire fungal body is segmented into spores.

Holomorph–Any fungus considered in its entirety, *i.e.*, including all latent or expressed (anamorphic or teleomorphic) forms.

Holotype–The single specimen designated or indicated as "the type" by the original author at the time of publication of the original description.

Homokaryon–A hyphal cell, mycelium, organism, or spore in which all the nuclei are genetically identical.

Homokaryotic–Refers to a homokaryon.

Homothallic–The condition of being self-fertile, able to reproduce sexually without a partner.

Horizontal resistance–In a given cultivar: the existence of similar levels of resistance to each of the races of a given pathogen.

Host–A plant that supports the growth and development of the parasite that has infected it.

Hyaline–Transparent, translucent, or colorless.

Hybrid–The offspring of two individuals differing in one or more heritable characteristics.

Hybridization–1. In molecular biology: The formation of stable duplexes between complementary sequences by way of Watson-Crick base-pairing.

2. Cross-breeding.

Hydathode–A specialized leaf structure with one or more openings through which water is discharged from the interior of the leaf to its surface.

Hydrosis–Necrotic symptom of disease characterized by water-soaking of tissues. (pl. **hydroses**.)

Hymenium–A palisade-like layer of asci or basidia, including any sterile cells, such as basidioles, paraphyses, or cystidia. (pl. **hymenia**.)

Hymenomycete–A member of the Hymenomycetes.

Hymenomycetes–The group of Basidiomycetes possessing an exposed hymenium.

Hymenophore–Spore-bearing structure; the part of a basidioma bearing the hymenium.

Hyperparasite–A parasite of a parasite.

Hyperplasia–The enlargement of an organ or tissue owing to an increase in the number of cells.

Hypersensitive–The state of being abnormally sensitive. It often refers to an extreme reaction to a pathogen (*e.g.*, the formation of local lesions by a virus or the necrotic response of a leaf to bacterial infection).

Hypersensitivity–The expression of extreme reactivity by a plant in response to a potential parasite or pathogen, the plant's response commonly serving to limit or prevent parasitization/disease.

Hypertrophy–Increase in cell size causing an increase in the size of an organ or tissue.

Hypha–Filamentous part of a fungus, usually septate and consisting of several cells in linear succession. (pl. **hyphae.**) Search in mycelium.

Hyphal fusion–Search in anastomosis.

Hyphal peg–1. A compound, hyphal, fasciculate projection extending beyond the general level of the hymenium, consisting of two or more parallel or interwoven hyphae, encrusted or gelatinized.

2. A projection from a hypha.

Hyphomycetes–A group of the Deuteromycetes without differentiated pycnidia or acervuli.

Hyphopodium–A short branch of one or two cells of the eplphytic mycelium of a black mildew fungus. (pl. **hyphopodia.**)

Hypodermis–In nematodes: A thin tissue layer beneath the cuticle that thickens to form the dorsal, lateral, and ventral chords, which extend the length of the body.

Hypodermal–Pertaining to the hypodermis.

Hypogeous–Growing below ground.

Hyponasty–More rapld growth of the lower side of an organ than of the upper side.

Hypoplasia–Underdevelopment resulting from an abnormal paucity of cells.

Hypovirulence–A reduced level of virulence in a strain of pathogen resulting from genetic changes in the pathogen or to the effects of an infectious agent on the pathogen.

Hysterothecium–An ascocarp that opens by a slit. (pl. **hysterothecia.**)

Idiomorphs–Nucleotide sequences that occupy the same locus in different strains but are not related in sequence or common descent.

Imbricate–Overlapplng, like the shingles of a roof.

Immune–Cannot be infected by a given pathogen.

Immunity–1. The state of being immune. 2. In plants, the ability to remain free from disease because of inherent structural or functional properties.

Imperfect fungus–Search in Deuteromycotina.

Imperfect stage–Search in anamorph.

Imperfect state–The state of a fungus characterized by asexual spores (conidia) or the absence of spores. Also called **imperfect stage.**

Incipient–Early in development (of a disease or condition).

Incisure–In nematodes: A longitudinal cuticular cleft that divides the lateral fields; sometimes called involution or line.

Incompatible–Not cross-fertile.

Incubation period–The period of time between penetration of a host by a pathogen and the first appearance of symptoms on the host.

Indeterminate–Without definite margin or edge; terminal growth (of hyphae or conidiophores) unrestricted.

Indexing–A procedure to determine whether a given plant is infected by a virus. It involves the transfer of a bud, scion, sap etc. from one plant to one or more kinds of indicator plants sensitive to the virus.

Indicator host–A plant species that gives characteristic symptoms to a specific virus. Used in virus diagnosis.

Infect–Of a parasite: To begin or continue an interactive, usually pathogenic, relationship with the host.

Infection–The interaction of parasite with host; the beginning of that interaction.

Infection court–A site in or on a host plant where infection can occur.

Infection thread–The specialized hypha of a pathogenic fungus that invades tissue of the susceptible plant.

Infectious–Capable of producing propagules that disperse from one host to another and infect it.

Infective–Of an agent of inoculation, capable of transmitting inoculum.

Infest–Of microorganisms and viruses: To contaminate the surface of a plant without establishing an interactive relationship with it; to be present in high numbers in a plant's environment (soil, water, etc.).

Ingress–The act, by a plant pathogen, of gaining entrance into the tissues of a susceptible plant.

Inoculate–To introduce a micro-organism into an environment suitable for its growth; to bring a parasite into contact with a host.

Inoculation–The act of inoculating; the placement of microorganisms or viruses at a site where infection is possible (the infection court).

Inoculum–The population of microorganisms introduced in an inoculation; the units of a parasite capable of initiating an infection.

Inoperculate–Of an ascus or a sporangium, opening by a pore or split to discharge spores, as in asci of the Helotiales.

Intercalary–Formed or situated somewhere between apex and base of a given structure.

Intercellular–Between cells.

Intersex–An individual more or less intermediate in phenotype between male and female, displaying secondary male or female characters.

Intracellular–Within or through the cells.

Intumescence–Hyperplastic symptom characterized by blister-like swelling on the surfaces of plant organs.

Invagination–Retraction, under force of pressure, of an outer surface toward the inside.

in vitro–(Literally "in glass".) Cultivated in an artificial, non-living environment.

in vivo–Within a living organism.

Invasion–In plant pathology: Spread of a pathogen through tissues of a diseased plant.

Involute–Of the margin of the plleus: Rolled in, especially when young.

Involution–Search in incisure.

Isoenzyme–Search in isozyme.

Isogamy–The condition in which gametes are morphologically similar, as in the members of the Zygomycotina.

Isogenic–Two or more organisms or cells having identical genotypes.

Isolate–In plant pathology: a culture or subpopulation of a microorganism separated from its parent population and maintained in some sort of controlled circumstance; also, to effect such separation and control, for example to isolate a pathogen from diseased plant tissue.

Isolation–1. The process of getting an organism in pure culture.

2. The pure culture itself.

Isoline–(Isogenic line.) One line in a series of genetically similar plant lines that carry different specific genes for resistance to a particular pathogen.

Isotype–Any duplicate of the holotype of a taxon.

Isozyme–One of a number of enzymes that catalyse the same reaction(s) but differ from each other in primary structure and/or electrophoretic mobility. (Also **isoenzyme**.)

Isthmus–In nematodes: Relatively narrow portion of esophagus just anterior to the basal region.

Juvenile–The life stage of a nematode between the embryo and the adult; an immature nematode.

Juvenillody–Condition in which tissues and organs remain immature.

Karyogamy–The fusion of nuclei.

Karyotype–The chromosomal constitution of a eukaryotic cell in terms of the number, size amd morphology of the chromosomes at metaphase.

Klendusity–A special kind of disease escape in which a susceptible plant avoids disease because of an intrinsic property of the plant itself that greatly reduces the chances of its being inoculated, even though there may be an abundance of inoculum in the area.

Knot–A localized abnormal swelling; a gall.

Labial–In nematodes: Pertaining to the lips.

Lamella–Search in gill. (pl. **lamellae**.)

Lateral field–In nematodes: An interruption of the transverse striae by longitudinal cuticular thickenings situated on top of the lateral chords. The field is divided by longitudinal striae (incisures) and at times by transverse markings (areolation).

Latent infection–1. Infection unaccompanied by visible symptoms. Search in **latent period**.

Latent period–1. The period between infection and the appearance of new inoculum. (Search in incubation period.)

2. Elapsed time between phage infection or induction and lysis of bacterial cells.

3. Period after acquisition of virus by a vector before it becomes infective.

Latent virus–A virus that infects a plant without causing macroscoplc symptoms.

Lateral organs–Search in amphid.

Leaf spot–A self-limiting lesion on a leaf is known as leaf spot.

Lectin–A plant protein that binds to certain sugar residues in glycoproteins.

Lectotype–One of a series of syntypes that, after publication of the original description, is selected and designated through publication to serve as the type.

Legitimate–Of taxonomic names and eplthets: Published in accordance with the Code of Nomenclature.

Leptoderan–In nematodes: Caudal algae that do not meet posterior to the tail tip.

Lesion–A localized area of diseased or damaged tissue.

Library–In molecular biology: A set of cloned fragments together representing the entire genome.

Lichen–A thallus consisting of an alga and fungus intermixed and living in a symbiotic relationship.

Life cycle–The complete succession of changes undergone by an organism during its life. A new cycle occurs when an identical succession of changes is initiated.

Line–Search in incisure.

Lips–In nematodes: Cuticular structures (usually six: two subdorsal, two lateral, two subventral) surrounding the mouth opening; lips may be fused in pairs.

Local infection–An infection affecting a limited part of a plant.

Local lesion–A localized spot produced on a leaf upon mechanical inoculation with a virus.

Locular–Containing chambers or hollows.

Locule–A cavity.

Loculoascomycete–A member of the Loculoascomycetes.

Loculoascomycetes–A group of Ascomycotina with bitunicately discharging asci, producing ascospores that are generally septate and borne in unwalled locules (pseudothecia) in ascostromatic ascomata with an ascolocular ontogeny. Not accepted by some authors.

Locus–Site on a chromosome occupied by a particular gene.

Lumen–In nematodes. Triradiate canal or duct of the esophagus. In fungi: The space bounded by tissue or wall, as the central cavity of a cell.

Lunate–Crescent-shaped; half-moon shaped.

Luteovirus–(Literally "yellowish".) Member of a group of plant viruses with isometric particles containing one molecule of linear RNA, mainly confined to the phloem, and usually not mechanically transmitted but transmitted in nature by aphids in a circulative manner.

Lyophilization–Rapid freezing of a material at low temperature followed by rapid dehydration by sublimation in a high vacuum. A method used to preserve biological specimens or to concentrate macromolecules with little or no loss of activity. (Also **freeze-drying.**)

Macroconidium–The larger of two types of conidia formed by certain fungi. (pl. **macroconidia.**) Search in microconidium.

Macrocyclic–Of rusts: The two primary spore stages are present (telial with teliospores and aecial with aeciospores).

Macroscoplc–Visible to the unaided eye.

Mammillate–Digitate, with nipple-shaped protuberances.

Masked symptoms–Virus-induced plant symptoms that are absent under some environmental conditions but appear when the host is exposed to certain conditions of light and temperature.

Mechanical inoculation–Of plant viruses, a method of experimentally transmitting the pathogen from plant to plant; juice from diseased plants is rubbed on test-plant leaves that usually have been dusted with carborundum or some other abrasive material.

Medium–In microbiology: any liquid or solid preparation made specifically for the growth, storage, or transport of microorganisms or other types of cell. (pl. **media.**)

Median bulb–Search in metacorpus.

Medulla–Central part of an organ.

Meiosis–The process in which a eukaryotic nucleus divides into nuclei whose ploidy is lower than that of the parent nucleus (typically, haploid nuclei being formed from diploid nuclei) and in which recombination usually occurs.

Meiospore–A uninucleate, haploid spore arising directly by meiosis.

Messenger RNA–A chain of specific ribonucleotides that codes for a specific protein; template for the assembly of amino acids into protein; in cells mRNA is transcribed from DNA, but some RNA viruses function directly as mRNA.

Metabasidium–The cell in which meiosis occurs in members of the Basidiomycotina. (Pl. **metabasidia.**)

Metacorpus–The swollen posterior portion of the corpus; sometimes called the median bulb.

Metaplasia–1.Changed condition of a structure or organ;

2.hyperplastic class of symptoms characterized by overdevelopment other than that due to hypertrophy or hyperplasia. (*e.g.*, abnormal starch accumulation, virescence, etc.).

Microconidium–1. A spermatium. 2. A small conidium; pl. **micro-conidia.**) Search in macroconidium.

Microcyclic–A life cycle in the rusts where one or more of the main spore stages, usually the aecial, is absent.

Micropyle–In nematodes: A minute opening in the membrane of an egg through which the spermatozoa enter.

Microsclerotium–A small clump of dark-colored, more or less thick-walled cells, each of which is viable; produced in culture and rarely in the xylem of host plants. (Search in sclerotium.) (pl. **microsclerotia**.)

Microscopic–Very small; search in only with the aid of a microscope.

Mildew–A fungal disease of plants in which the mycelium and spores of the fungus are search inn as a whitish growth on the host surface.

Minimal medium–A type of culture medium lacking specific growth factors; it does not support the growth of some or all auxotrophic strains of a given organism but permits the growth of prototrophic strains.

Mitosis–A sequence of cellular events that culminates in the division of a eukaryotic nucleus into two genetically similar or identical nuclei whose ploidy is the same as that of the parent nucleus. Mitosis occurs during asexual cell division.. (pl. **mitoses**.)

Mitospore–A uninucleate, haploid or diploid spore arising by mitosis.

Mitosporic fungi–Search in Deuteromycotina.

MLO–Mycoplasma-like organism. Search in phytoplasma.

Mold–A downy fungal growth on a substratum, usually consisting of mycelium of a Hyphomycete or a Zygomycete.

Mollicute–A proposed trivial name for any member of the class Mollicutes. The use of this name could permit "mycoplasmas" to be used specifically for members of the genus Mycoplasma.

Molt–To cast off the cuticle.

Monocyclic–Of a disease or pathogen: Producing one generation of inoculum and one cycle of infection during a single growing season. (Search in polycyclic.)

Monodelphic–Of nematodes: Possessing one genital tube or ovary.

Monoecious–Having male and female reproductive organs on a single thallus.

Monogenic–Of nematodes: Producing offspring of only one sex.

Monogenic resistance–Resistance determined by a single gene.

Monokaryotic–Having one nucleus per cell.

Monophialide–Search in conidiophore.

Monotype–The sole species of a newly proposed genus.

Morphologic–Pertaining to form.

Mosaic–A common symptom induced in leaves by many plant virus infections in which there is a pattern of dark green, light green and sometimes chlorotic areas. This pattern is often associated with the distribution of veins in the leaf. In monocotyledonous leaves it shows as stripes.

Mottle–A diffuse form of the mosaic symptom in plant leaves in which the dark and light green are less sharply defined. This term is frequently used interchangeably with mosaic.

mRNA–Search in messenger RNA.

Mucro–In nematodes: A stiff or sharp point abruptly terminating an organ.

Mucronate–In nematodes: Ending in a sharp point.

Mulch–A protective covering that is spread on the ground around plants to inhibit evaporation and weed growth, control soil temperature, enrich the soil, or prevent the dispersal of pathogens. It may be organic material such as leaves, peat, or wood chips, or inorganic material such as plastic sheeting.

Multicomponent virus–A virus in which the genome needed for full infection is divided between two or more particles (*e.g.*, cowpea mosaic virus, brome mosaic virus, cucumber mosaic virus).

Multiline–A plant cultivar made up of a mixture of isolines differing in their genes for resistance to a particular pathogen. (Also **multiline cultivar.**)

Multiseptate–Having more than one septum.

Mummy–A dried, shrivelled fruit colonized by a fungus.

Muriform–Having bricklike cells in a wall with both longitudinal and transverse septa.

Mushroom–A fleshy fruiting body of a fungus, especially of a basidiomycete of the family Agaricaceae.

Mutant–Of an organism, population, gene, chromosome, etc.: Differing from the corresponding wild type by changes in one or more loci.

Mutation–A stable, heritable change in the nucleotide sequence of a genetic nucleic acid (DNA, or RNA in viruses, viroids, etc) typlcally resulting in the generation of a new allele and a new phenotype.

Mutualism–Search in symbiosis.

Mycelium–A mass of hyphae, often used to denote all hyphae comprising a thallus. (pl. **mycelia.**)

Mycology–The study of fungi.

Mycoparasite–A fungus parasitic on other fungi.

Mycoparasitism–One fungus living on another.

Mycophagous–Feeding on fungi.

Mycoplasma–A genus of cell wall-less, sterol-requirlng, catalase-negative bacteria (family Mycoplasmataceae) occurring as parasites and pathogens.

Mycoplasma-like organism–Search in phytoplasma.

Mycorrhiza–A specialized root structure resulting from a symbiotic relationship between a fungus and a higher plant. (pl. **mycorrhizae.**)

Mycosis–Disease in animals caused by a fungus. (pl. **mycoses.**)

Mycotoxicosis–Any disease of man or animals resulting from the ingestion of mycotoxins.

Mycotoxin–A toxin produced by a fungus. The term is usually reserved for fungal metabolites that are toxic to man and/or animals and are produced by molds growing on foodstuffs (*e.g.*, aflatoxins, ergot alkaloids).

Mycotrophic–Refers to green plants having mycorrhizae.

Mycovirus–A virus that replicates in cells of fungi.

Myxamoeba–A naked cell capable of amoeboid movement; characteristic of the vegetative phase of myxomycetes and such Plasmodiophoromycetes as *Plasmodiophora brassicae*.

Myxomycetes–Search in **Myxomycota**.

Myxomycota–The slime molds, a class of fungi characterized by amoeboid vegetative protoplasts, plasmodia, and by brightly coloured spore bearing caplllitia.

Necrosis–Localized death of cells or tissues.

Necrotic–Dead.

Necrotroph–1. An organism that kills part or all of another organism before deriving nutrients from it (usually applied to plant pathogens). 2. An organism that derives nutrients from dead plant or animal tissues, whether or not it is responsible for the death of those tissues.

Nematicide–A chemical compound or physical agent that kills nematodes.

Nematode–More or less elongate, splndle-shaped, worm-like animals ranging in size from less than a millimeter to several meters in length, living as saprophytes in soil or water or as parasites of plants or animals.

Neotype–A specimen selected as the type subsequent to the original description in cases in which the primary types are definitely known to have been destroyed.

Nepovirus–(Siglum of nematode polyhedral virus). Member of a group of multicomponent plant viruses with two isometric particles containing two species of linear RNA, transmitted mechanically and by soil-inhabiting nematodes.

Nerve ring–The center of the nervous system of nematodes that encircles the esophagus; composed largely of nerve fibers and associated ganglia.

Nomenclature–A system of names, or naming, as applied to the subjects or study in any art or science, especially in botany and zoology.

Noncirculative transmission–Virus transmission characterized by a very short period of acquisition of the virus by a vector (*e.g.*, an aphid), no latent period before the vector can transmit the virus, and a short period of retention by the vector after acquisition. (Also termed **non-presistent transmission.**)

Nonseptate–Lacking cross-walls.

Northern blot–An RNA blot. This term originated as lab jargon; not a generally accepted term.

Nuclease–Any enzyme that can cleave the sugar-phosphate backbone of a nucleic acid.

Nucleoprotein–A compound of nucleic acid and protein.

Nutrient broth–A liquid basal medium.

Obligate–Restricted to a particular set of environmental conditions, without which an organism cannot survive. (*e.g.*, an obligate parasite can survive only by parasitizing another organism.)

Obligate anaerobe–An organism that can grow only under anaerobic conditions.

Obligate parasite–An organism that is incapable of living as a saprophyte and must live as a parasite.

Oblong–(Of spores) longer than broad (about twice as long or somewhat less), with sides nearly parallel and with ends more or less flattened.

Obovate–Egg-shaped, with the wide end outward.

Obovoid–Egg-shaped, with the narrow end outward.

Obtuse–(Of plleus, cystidia, spores) rounded or blunt.

Odontostylet–Search in stylet. Synonymous with onchiostylet.

Oedema–Intumescence or blister formation because of an increase in intercellular water. (Also **edema**.)

Ontogeny–Development of the individual.

Onchiostylet–In nematodes: A stylet developed from a special cell in the anterior part of the esophagus from which it moves into place during each molt.

Oocyst–Synonymous with oogonium.

Oocyte–Female germ cell.

Oogamy–A type of heterogamy in which plasmogamy takes place between a large nonmotile egg and a small motile male gamete or cytoplasm from an antheridium.

Oogonium–A female gametangium that contains one or more discrete gametes. (pl. **oogonia**.)

Oomycetes–A class of aquatic and terrestrial fungi (subdivision Mastigomycotina) that typlcally produce oogonia and zoosporangia in which form zoospores having one anteriorly-directed tinsel flagellum and one posteriorly-directed whiplash flagellum.

Oomycete–A member of the Oomycetes.

Oosphere–A large, naked, nonmotile, usually spherical, female gamete.

Oospore–A thick-walled spore that develops from an oosphere through plasmogamy or parthenogenesis.

Operculate–Of an ascus or sporangium, opening by a subcircular aplcal lid to discharge spores, as in asci of the Pezizales.

Operculum–A flap or lid-like covering over the opening of an ascus or sporangium. (pl. **opercula**.)

Opisthodelphic–Of nematodes: Having the uterus (or uteri) directed posteriorly.

Ostiole–1. A neck-like structure in an ascocarp, lined with pariphyses, and terminating in a pore. 2. The opening of a pycnidium.

Ostiolum–Search in ostiole. (pl. **ostiola.**)

Ovary–Female sexual gland in which the ova, or eggs, are formed.

Oviduct–A short, usually tubular, thick-walled part of the female reproductive system between the ovary and spermatheca.

Oviparous–Producing eggs that hatch after expulsion from the body.

Ovoid–Egg-shaped.

Ovoviviparous–Producing eggs that hatch within the body.

PAGE–Polyacrylamide gel electrophoresis.

Papilla–A hump or swelling. (pl. **papillae.**)

Papillate–Bearing a papllla.

Paraphysis–Sterile, elongated cell that may occur in the hymenium, intermixed with asci or basidia, elongating apically and having a free apex. (pl. **paraphyses.**)

Parasexual cycle–A sequence involving heterokaryon formation, diploidization, and haploidization, often resulting in the formation of recombinant nuclei. Unlike the sexual cycle, the parasexual cycle can occur at any point or continuously throughout the life cycle.

Parasite–An organism living in or on another living organism (host) from which it extracts nutrients.

Paratype–A specimen other than the holotype and its isotypes that the author cited at the time of publication of the original description.

Parasitic–Having the characteristics of a parasite.

Parenchyma–A tissue composed of living, thin-walled cells that can continue to divide even when mature; parenchyma cells usually leave intercellular spaces between them.

Parthenogenesis–Process of reproduction by the development of an unfertilized egg.

Parthenogenic–Pertaining to parthenogenesis.

Pathogen–An agent (biotic or abiotic) that causes plant disease. **pathogenesis.** That portion of the life cycle of a pathogen during which it becomes, and continues to be, associated with its suscept.

Pathogenic–Having the characteristics of a pathogen.

Pathogenicity–The capability of a pathogen to cause disease.

Pathology–1. The study of disease. 2. The abnormal condition that constitutes disease.

Pathotype–An infrasubspecific classification of a pathogen distinguished from others of the species by its pathogenicity on a specific host(s).

Pathovar–In bacteria: An infrasubspecific group that can infect only plants within a certain genus or species. Search in pathotype.

PCR–Search in polymerase chain reaction.

Pellet–The material concentrated at the bottom of a centrifuge tube after centrifugation.

Peloderan–In nematodes: Caudal alae that meet posterior to the tail tip.

Penetration peg–In some plant parasitic fungi: The peg-like hypha emerging from an appressorium that penetrates the epldermal cell wall.

Perfect state–The state of a fungus characterized by sexual spores.

Peridermium–A blister-like aecium as in the form genus *Peridermium*. (pl. **peridermia**.)

Peridium–A wall or membrane of sterile cells around a fruiting body (*e.g.*, around a sporangium or delimiting an aecium). (pl. **peridia**.)

Perineal pattern–Fingerprint-like pattern formed by cuticular striae surrounding the vulva and anus of the mature *Meloidogyne* female.

Periphysis–Short, hair-like filaments that line the canal of the ostiole in some Pyrenomycetes. (pl. **periphyses**.)

Perithecioid–Like a perithecium.

Perithecium–A closed ascocarp with a pore at the top, a true ostiole, and a wall of its own. (pl. **perithecia**.)

Peritrichous–Of bacterial flagella: Distributed more or less uniformly over the cell surface.

Peronosporales–Specialized forms of the Oomycetes, including aquatic and terrestrial species; many species in this order are plant pathogens (damplng-off fungi, downy mildews, and white rusts); unlike the true fungi, they lack chitin in their cell walls.

Persistent transmission–Search in circulative transmission.

Phage–A general term used for viruses isolated from prokaryotes including bacteria, blue-green algae (cyanobacteria) and mollicutes (phytoplasma and spiroplasma). The viruses from these different host groups are termed bacteriophages, cyanophages and mycoplasmaphages, respectively.

Phasmid–In nematodes: A pore-like structure located in the lateral field of the posterior region of nematodes belonging to the class Secernentea. Function is believed to be sensory. Sometimes called precaudal glands.

Phenotype–The observable characteristics of an organism, either in total or with respect to one or more particular named characteristics.

Phloem–Food-conducting tissue, consisting of sieve tubes, companion cells, phloem parenchyma, and fibers.

Phragmobasidium–A basidium that is divided into more than one cell by transverse or longitudinal septa. (pl. **phragmobasidia**.)

Phragmospore–A spore having two to many transverse septa.

Phragmosporous–Having transversely multiseptate spores.

Phyllody–A change of floral petals (leaves) to foliage leaves.

Phylloplane–The surface(s) of a leaf.

Physiological race–A subdivision of a species of pathogen, particularly fungi, distinguished from other members of the species by specialization for pathogenicity to different cultivars of a host.

Physiologic specialization–The existence of a number of races or forms of one species of pathogen based on their pathogenicity to different cultivars of a host.

Phytoalexin–A low molecular weight, antimicrobial compound synthesized by and accumulating in higher plants exposed to certain microorganisms (pathogenic and nonpathogenic).

Phytopathogenic–Of microorganisms: Capable of initiating disease in plants.

Phytoplasma–A prokaryotic, plant parasitic microorganism resembling a mycoplasma but not yet isolable in pure culture or characterized taxonomically.

Phytotoxic–Toxic to plants.

Phytotoxin–1. A toxin produced by a microorganism and active against a plant or against plant cells/tissues. 2. A toxin produced by a plant.

Pileate–Possessing a cap or plleus.

Pileus–The expanded caplike portion of some basidiocarps or ascocarps that supports the hymenium. (pl. **pilei**.)

Plasmalemma–The cytoplasmic membrane found on the outside of the protoplast adjacent to the cell wall.

Plasmid–In many types of prokaryotic and eukaryotic cell: a linear or covalently closed circular molecule of DNA, (distinct from chromosomal DNA, mtDNA, ctDNA, or kDNA and commonly dispensable to the cell), that can replicate autonomously (*i.e.*, independently of other replicons).

Plasmodesma–A fine protoplasmic thread connecting two protoplasts and passing through the wall separating the two protoplasts. (pl. **plasmodesmata**.)

Plasmodium–A multinucleated, usually naked (*i.e.*, bounded only by a plasma membrane) mass of protoplasm that is usually motile and variable in size and form. (pl. **plasmodia**.)

Plasmogamy. The fusion of two protoplasts.

Plasmolysis-The shrinking and separation of the cytoplasm from the cell wall due to exosmosis of water from the protoplast.

Plectomycete–A member of the Plectomycetes.

Plectomycetes–In general, a group of primitive or reduced ascomycetous forms that have an angiocarpous fructification without an ostiole, the entire interior of which is irregularly penetrated by ascogenous hyphae, with the result that the generally spherical asci, without accompanying paraphyses or other threads, lie scattered irregularly in a pseudoparenchymatous tissue composed of the ascogenous hyphae.

Pleomorphic–Exhibiting pleomorphism.

Pleomorphism–1. In fungi: Having more than one independent form or spore stage in the life cycle.

2. In general: An inherent variability in size and shape (*e.g.*, among the

cells in a pure culture or clone of a given organism).

Plerome–The plant tissues inside the cortex.

Plesionecrosis–A symptom exhibited by tissues not yet dead but in the process of dying (*e.g.*, wilting). (pl. **plesionecroses**.)

Ploidy–The number of (complete) sets of chromosomes in a cell.

Polar–At one end or pole of the cell (*e.g.*, a flagellum, spore inclusion, germ tube, etc.).

Polycyclic–Of a disease or pathogen: Producing many generations of inoculum and many cycles of infection during a single growing season. (Search in monocyclic.)

Polyetic–Of plant disease epidemics: Continuing from one growing season to the next.

Polymerase chain reaction–The selective amplification of DNA by repeated cycles of (a) heat denaturation of the DNA, (b) annealing of two oligonucleotide primers that flank the DNA segment to be amplified and (c) the extension of the annealed primers with the heat insensitive *Tag* DNA polymerase.

Polymorphism–Pleomorphism.

Polygenic–A character controlled by many genes.

Polynucleate–Having more than one nucleus per cell.

Polypore–A member of the Polyporaceae; the hymenium forms tubes in the basidiocarp.

Poroconidium–Search in porospore.

Porospore–A conidium produced by an extension of the inner wall of the conidiogenous cell and extrusion through a pore in the wall of the conidiophore.

Potexvirus–(Siglum of potato virus X.) Member of a group of plant viruses that infect a wide range of hosts, including monocots and dicots; individual members infect only a narrow host range, typically causing mosaic and ringspot symptoms. Type member: potato virus X (PVX).

Potyvirus–(Siglum of potato virus Y). Member of a large group of plant viruses with flexuous particles containing a single molecule of linear RNA, most of which are transmitted by aphids in a noncirculative manner.

Predisposition–An increase in susceptibility resulting from the influence of environment on the suscept.

Primary cycle–Of plant disease: the first infection cycle to begin in a given season; usually occurring only once per season.

Primary infection–The first infection of a plant by a pathogen emerging from a dormant stage in its life cycle (overwintering or oversummering).

Primary inoculum–The overwintering or oversummering pathogen or its propagules that cause primary infection.

Probasidium–The cell in which karyogamy occurs in the basidiomycetes. (pl. **probasidia**.)

Probe–A specific sequence of DNA or RNA used to detect complementary sequences by hybridization.

Procorpus–In nematodes: Cylindrical portion of the corpus anterior to the metacorpus.

Prodelphic–In nematodes: Having uteri parallel and anteriorly directed at the origin.

Proliferation–A rapld and repeated production of new cells, tissues, or organs; specifically, a hyperplastic symptom of plant disease in which organs continue to develop after they have reached the point beyond which they normally do not grow.

Prolepsis–A hyperplastic symptom of disease in which organs appear before the natural time. (pl. **prolepses**.)

Promoter–A region of DNA, usually upstream of a coding sequence, that binds RNA polymerase and directs the enzyme to the correct transcriptional start site.

Promycelium–The short hypha bearing sporidia produced by the teliospore; the basidium. (pl. **promycelia**.)

Propagative virus–A circulative virus that replicates in its insect vector. Such a virus is said to be **propagatively transmitted** (*e.g.*, potato yellow dwarf virus).

Propagule–Any disseminative unit of an organism (*e.g.*, a spore, a mycellal fragment a sclerotium).

Protease–A generic term for an enzyme that cleaves a polypeptide chain.

Protectant–Any chemical agent that interacts with a pathogen on the plant surface to inhibit infection before it takes place. Non systemic.

Protection–A principle of plant disease control in which a barrier is placed between suscept and pathogen (*e.g.*, the use of protective chemical dusts or sprays).

Prototroph–A strain of microorganism whose nutritional requirements do not exceed those of the corresponding wild-type strain.

Protoplast–A plant cell from which the cell wall has been removed.

Pseudocoel–In nematodes: Body cavity containing a fluid in which the various internal organs are suspended.

Pseudoparaphysis–A sterile thread that grows downward in the cavity of some ascocarps, usually becoming attached at the bottom. (pl. **pseudoparaphyses**.)

Pseudothecium–An ascostroma resembling a flask-shaped perithecium. (pl. **pseudothecia**.)

Punctate–Having surface dot(s), pore(s), etc.

Pustule–Small blister-like elevation of the leaf epldermis created as spores emerge from underneath and push outward.

pv–Pathovar.

Pycnidiospore–A conidium formed in a pycnidium.

Pycnidium–A closed sporocarp, usually opening by a pore, that contains a cavity bearing conidia. (pl. **pycnidia**.)

Pycniospore–A spore (spermatium) borne in a pycnium in the Uredinales.

Pycnium–In the Uredinales, stage 0, consisting of the male fertilizing elements (pycniospores) and female elements, the flexuous hyphae. (pl. pycnia.)

Pycnospore–An obsolete term for a pycnidiospore or a pycniospore.

Pygmism–The state of being dwarfed or reduced in size.

Pyrenomycete–One of the Pyrenomycetes.

Pyrenomycetes–Class of Ascomycotina, traditionally based on taxa with perithecioid ascomata that are ascohymenial in ontogeny and have unitunicate asci, often with an aplcal annulus.

Pyriform–Pear-shaped.

Quarantine–Legal restriction of the movement of plant pests (or the products that may be harboring them) into areas where they do not occur.

Quiescence–That period of the prepenetration stage during which a pathogen may be inactive bccause environmental conditions are unfavorable for its growth.

Race–A subspecies group of pathogens that infect a given set of plant varieties.

Rachis–In fungi: A conidiophore elongating to one side of a terminally produced spore, often resulting in a zig-zag-shaped structure.

Range–Of a plant pathogen: The geographical region or regions in which it is known to occur.

Recombinant DNA–DNA molecules in which sequences, not normally contiguous, have been placed next to each other by *in vitro* methods.

Rectum–In nematodes: Posterior gut of the female. A narrow, dorso-ventrally flattened tube that is lined with cuticle and separated from the intestine by a sphincter muscle.

Reflexed–Bent back.

REMI (Restriction enzyme mediated integration)–A method of transformation that generates tagged mutations.

Reniform–Kidney-shaped.

Replicon–Any DNA sequence or molecule that possesses a replication origin and is therefore potentially capable of being replicated in a suitable cell.

Resistance–The ability of an organism to exclude or overcome, completely or in some degree, the effect of a pathogen or other damaging factor.

Resistant-Possessing resitance.

Response–The change produced in an organism by a stimulus.

Resting spore–A thick-walled spore, particularly one formed by a sexual process, germinating only after an extended period of dormancy (*e.g.*, an overwintering teliospore).

Restriction endonuclease–An endo-nuclease that binds to double-stranded DNA at a specific nucleotide sequence and then, if both strands of the DNA lack appropriate modification at that sequence, cleaves the DNA either at the recognition sequence or at another site in the DNA molecule.

Reticulate–Having net-like markings.

Retrorse–In a backward or downward direction.

Revolute–Of the margin of the plleus: rolled back or up.

RFLP (Restriction fragment length polymorphism)–Inherited differences in sites for restriction enzymes (*e.g.*, caused by base changes in the target site) that result in differences in the lengths of the fragments produced by cleavage with the relevant restriction enzyme. RFLPs are used for genetic mapplng to link the genome directly to a conventional genetic marker.

Rhabdions–In nematodes: Plates in the cuticular lining of the stoma that make up the walls of the various divisions of the stoma.

Rhizoid–A root-like structure forming part of the thallus in certain algae and fungi; it may anchor the organism to the substratum and/or act as an absorptive organ.

Rhizomorph–A macroscoplc, typlcally rope-like strand of compacted tissue formed by certain higher fungi. Rhizomorphs often are enduring structures that can remain dormant under adverse conditions.

Rrhizomycelium–Branching, anucleate or sparsely nucleate, rhizoidal filaments of variable width forming part of the thallus in some fungi. (pl. **rhizomycelia**.)

Rhizoplane–The root surface.

Rhizosphere–An environment regarded, variously, as (a) that region of the soil modified as a result of the uptake and deposition of substances by a growing root, (b) the root itself, together with that volume of soil which it influences, (c) the root surface together with that region of the surrounding soil in which the microbial population is affected by the presence of a root.

Rickettsiae–Microorganisms similar to bacteria in most respects but generally capable of multiplying only inside living host cells; parasitic or symbiotic.

Ringspot–A type of local lesion consisting of single or concentric rings of discoloration or necrosis, the regions between the concentric rings being green. The center of the lesion may be chlorotic or necrotic.

RNA blotting–A technique for transferring RNA from an agarose gel to a nitrocellulose filter on which it can be hybridized to a complementary DNA.

Roestelium–In the Uredinales an aecium with a cornute peridium, thin at the sides, usually rupturing by longitudinal slits, and made up of characteristically marked and imbricated cells. The name is taken from the form-genus *Roestelia*. (pl. **roestelia**.)

Rogue–A variation from the standard varietal type; also, to remove such undesirable plants (especially those infected with viruses) from the growing crop.

Roguing–The removal of diseased plants from a crop in order to prevent the spread of the disease.

Rosette–An abnormal condition in which the leaves form a radial cluster on the stem.

Rot–The softening, discoloration, and often disintegration of a succulent plant tissue as a result of fungal or bacterial infection.

Roundworm–Nematode.

Rugose–Wrinkled.

Russet–Brownish, roughened areas on the skin of fruit as a result of cork formation.

Rust–1. Fungus of the class Urediniomycetes. 2. Any of various plant diseases caused by members of the Urediniomycetes or by species of *Albugo*. The diseases are called "rusts" because many of the causal agents form rust-coloured spores on affected plants.

Saccate–Pouch–or sac-like.

Saltation–A mutation occurring in the asexual state of fungal growth, especially one occurring in culture.

Sanitation–Principle of plant disease control involving removal and burning of infected plant parts and decontamination of tools, equipment, hands, etc.

Saprobe–An organism that obtains its nutrients from non-living organic matter (commonly dead and decaying plant or animal matter) by absorbing soluble organic compounds. (Also **saprotroph**.)

Saprogenesis–Survival; that phase of the life cycle of a pathogen during which it is not actively causing disease in a living suscept.

Saprophyte–Search in saprobe.

Saprotroph–Search in saprobe.

Sarcody–It is a hyperplastic symptom in which swellings occur above and below portions of organs that are tightly encircled.

Satellite virus–A defective virus requiring a helper virus to provide functions necessary for replication is said to be a satellite virus. It may code for its own coat protein or various other products.

Scab–Any of a wide range of unrelated plant diseases having a roughened, crust like diseased area on the surface of a plant organ (*e.g.*, apple scab, potato scab, wheat scab).

Scabrous–Rough with short, rigid projections.

Scald–A necrotic condition in which tissue is usually bleached and has the appearance of having been exposed to high temperatures.

Sclerotium–Hard, resistant, multicellular resting body, usually with a differentiated cortex and medulla, that under favorable conditions can germinate to produce mycelium or sexual or asexual fruiting bodies. (pl. **sclerotia**.)

Sclerotized–In nematodes: Hardened refractive regions.

Scolecospore–A very long, thin spore (with a length/width ratio more than 15:1).

Scolecosporous–Having long, thin (filiform) spores.

Scorch–"Burning" of leaf margins as a result of infection or unfavorable environmental conditions.

Scutellum–In nematodes: An enlarged, shield-like phasmid. (pl. **scutella**.)

Secondary cycle–Of plant disease: any cycle initiated by inoculum generated during the same season.

Secondary infection–Any infection caused by inoculum produced as a result of a primary or a subsequent infection; an infection caused by secondary inoculum.

Secondary inoculum–Inoculum produced by infections that took place during the same growing season.

Secondary organism–An organism that multiplies in already diseased tissue but is not the primary pathogen, is known as secondary organism.

Secondary rot–Rot caused by a secondary organism.

Secondary symptom–A symptom of virus infection appearing after the first (primary) symptoms.

Sedentary–Staying in one place; stationary.

Senescence–Decline or degeneration, as with maturation, age, or disease stress.

Senescent–Aged, degenerate.

sensu lato–In a broad sense.

sensu stricto–In a narrow sense.

Septate–Having cross walls.

Septum–A cross wall in a hypha or spore. (pl. **septa**.)

Sequence–The order of nucleotides in RNA or DNA or of amino acids in a polypeptide.

Sequencing–Determining a sequence of a nucleic acid or protein.

Serology–Branch of science dealing with properties and reactions of sera, particularly the use of antibodies in the sera to examine the properties of antigens.

Serotype–A subdivision of virus strains distinguished by protein or a protein component that determines its antigenic specificity.

Seta–1. In nematodes: Elongated cuticular structures articulating with the cuticle; in general, tactile sensory organs usually located around the oral openings. In fungi: A bristle-like structure in some types of fruiting bodies. (pl. **setae**.)

Setose–Bristly; beset with bristles.

Sexual dimorphism–A pronounced difference in the morphologies of the two sexes within a species.

Serum–The fluid fraction of coagulated (clotted) blood.

Shock symptoms–The severe, often necrotic symptoms produced on the first new growth following infection with some viruses. Also called **acute symptoms**.

Shot hole–A symptom in which small, diseased fragments of leaves fall off, leaving small holes in their place.

Sieve plate–Perforated wall area between two phloem sieve cells through which they are connecting sieve tube protoplasts.

Sieve tube–A series of phloem cells forming a long cellular tube through which food materials are transported.

Sigmoid–Doubly curved in opposite directions, like the Greek letter sigma.

Sign–A visible manifestation of a causal agent of plant disease (*e.g.*, fungal spores or other fungal structures, bacterial ooze).

Slime mold–A member of a category of eukaryotic organisms that typlcally have some fungal-like attributes and some animal-like attributes.

Smut–Any of a number of plant diseases caused by the smut fungi (Ustilaginales); characterized by masses of dark, powdery, and sometimes odorous spores (*e.g.*, stinking smut of wheat, common smut of maize).

Smut spore–A dark, thick-walled resting spore of a smut fungus; may germinate to produce a promycelium; often improperly termed a chlamydospore.

Somatic cell hybridization–Production of hybrid cells by fusion of two protoplasts with different genetic makeup.

Somaclonal variation–Variability in clones generated from a single mother plant, leaf, etc. by tissue culture.

Sooty mold–A fungus of the family Capnodiaceae and of certain other families of the order Dothideales. The organisms grow eplphytically, utilizing honeydew, and form dark, spongy, hyphal mats on the surfaces of certain plants.

Sorus–It is a cluster or mass of spores or sporangia. (pl. **sori.**)

Southern blotting–Procedure for transferring denatured DNA from an agarose gel to a nitrocellulose filter where it can be hybridized with a complementary nucleic acid.

Spear–Search in stylet.

Species–The basic category of biological classification, displaying a high degree of mutual similarity determined by a consensus of informed oplnion; a subcategory of genus.

Species name–A Latin name consisting of two words, the generic name and the species eplthet (*e.g., Puccinia graminis*).

Species epithet–The second word of a species name (*e.g., graminis* in the species *Puccinia graminis*).

Spermagonium–In certain fungi: a structure within which male reproductive cells (spermatia) are formed. In rust fungi, a spermagonium is called a pycnium. (pl. **spermagonia.**)

Spermatheca–In nematodes: An enlarged portion of the female gonad between the oviduct and the uterus functioning in the storage of the sperm.

Spermatium–1. A non-motile male reproductive cell that can function in spermatization. 2. The male gamete of the rust fungi. Search in pycniospore. (pl. **spermatia.**)

Spermatization–In certain higher fungi: the union of a spermatium with a female reproductive structure.

Spermatocyte–In nematodes: A cell giving rise to spermatozoa or spermatozooids.

Spicate-Having the form of a splke.

Spicule–Male copulatory organ. Sometimes called **spiculum**.

Spiroplasma–A member of a group of pleomorphic, wall-less prokaryotes occurring as eplphytes or as intracellular or extracellular parasites or pathogens in a range of invertebrates and plants.

Sporangiophore–A modified hypha that supports the sporangium.

Sporangiospore–An asexual spore produced within a sporangium.

Sporangium–A sac that bears endogenous, asexual spores (sporangiospores) (pl. **sporangia**.)

Spore–A discrete sexual or asexual reproductive unit, usually enclosed by a rigid wall, capable of being disseminated.

Sporidiole–A little spore. (Also **sporidiolum**.)

Sporidiolum–Search in sporidiole.

Sporidium–1. A basidiospore formed by the rust or smut fungi. 2. In smut fungi: a spore formed in germination by repetition. (pl. **sporidia**.)

Sporocar–A fruit body that produces spores.

Sporodochium–A cluster of conidiophores arising from a stroma or mass of hyphae. (pl. **sporodochia**.)

Sporophore–A spore-producing or spore-bearing structure such as a conidiophore, ascocarp, or basidiocarp.

Sporulate–To produce spores.

Sporulation–The process of producing spores.

Spot–A symptom of disease characterized by a limited necrotic area, as on leaves, flowers, and stems.

Spreader–A substance added to fungicide or bactericide preparations to improve contact between the spray and the sprayed surface; a surfactant.

Squamous–Covered with or consisting of scales.

Squamule–A small scale or lobe.

Staurospore–A non-septate or septate spore with more than one axis.

Staurosporous–Possessing stauro-spores.

Stem-pitting–A symptom of some viral diseases characterized by depressions on the stem of the plant.

Sterigma–A splke-like structure that supports a basidiospore on the basidium. (pl. **sterigmata**.)

Sterile–1. Unable to reproduce sexually. 2. To be free from living microorganisms.

Sterile fungus–A fungus that is not known to produce any kind of spores.

Sterilization–The elimination of pathogens and other living organisms from soil, containers, etc.

by means of heat, chemicals, or radiation.

Sterilized–Free from living microorganisms.

Sticker–Added to fungicide or bactericide preparations to improve the adhesion of the spray to the sprayed surface.

Stipe–The stalk-like portion of some larger ascocarps or basidiocarps; any spore-bearing stalk.

Stoma–1. In plants: Aperture in the epldermis of a leaf, stem, or fruit, bound by two guard cells and functioning in gas exchange. 2. In nematodes: Buccal capsule. (pl. **stomata**.)

Strain–A sub-species group of organisms distinguishable from the rest of the species by a heritable characteristic that the individuals in the group have in common.

Stomatostylet–Search in stylet.

Striae–In nematodes: Superficial grooves or clefts on the cuticle; when present, striae may be search inn encircling lips or body.

Striate–Marked with delicate lines, grooves, or ridges.

Stroma–A compact mass of vegetative tissue, sometimes intermixed with host tissue, often bearing sporocarps either within or upon its surface. (pl. **stromata**.)

Stylet–In nematodes: Hollow protrusible spear used to puncture plants or animal prey.

Stylet knob–One of the (usually three) basal protuberances of the stylet.

Stylospore–A spore borne on a filament or hypha.

Suberized–Of cell walls: Hardened by their conversion to cork (suberin).

Subsp–.Subspecies.

Subspecies–An infraspecific population defined on the basis of one or more characters (morphologic for most organisms) that distinguish its members from typical representatives of the species.

Substrate–1. The material or substance on which a microorganism feeds and develops. 2. a substance acted upon by an enzyme.

Supe rinfection–Attempt to infect a host with a second virus, usually a different strain of the first infecting virus.

Supernatant–The material remaining above the pellet after centrifugation of a suspension.

Supplements–In nematodes: Preanal genital papillae on the ventral side of males; derived from cuticle, but may be provided with glands. Function during copulation. Sometimes called supplementary organs.

Suppression–A hypoplastic symptom characterized by the failure of plant organs or substances to develop.

Suppressive soil–A soil in which certain disease(s) fail to develop because of the presence in the soil of microorganisms antagonistic to the pathogen.

Suscept–Any plant that can be attacked by a given pathogen; a host.

Susceptible–Lacking the inherent ability to resist disease or attack by a given pathogen; not immune.

Susceptibility–The inability of a plant to resist the effect of a pathogen or other damaging factor.

Suspensor–A specialized hyphal tip that supports a gametangium and eventually a zygosporangium (in the Mucorales).

Swarmspore–Zoospore.

Symbiont–One member of a symbiotic relationship or existence between two individuals where both derive benefit from each other.

Symbiosis–A mutually beneficial association of two or more different kinds of organisms.

Symbiote–Symbiont.

Symptom–A visible abnormality in a plant that results from disease.

Symptomatology–The study of symptoms of disease and signs of pathogens for the purpose of diagnosis.

Symptomless carrier–A plant that, although infected with a pathogen (usually a virus), produces no obvious symptoms.

Synanamorph–Any one of the two or more anamorphs that are formed by the same fungus.

Syncytium–A multinucleate cell. (pl. **syncytia.**)

Synergism–The concurrent parasitism of a host by two pathogens in which the symptoms or other effects produced are of greater magnitude than the sum of the effects of each pathogen acting alone.

Synnema–A fascicle of conidiophores, usually upright; a coremium. (pl. **synnemata.**)

Syntype–One of a number of specimens of equal nomenclatural rank that formed all or part of the material the original author had in cases in which the author did not designate or indicate a holotype.

Systemic–Of a chemical or a pathogen: Spreading internally throughout the plant body.

Systemic infection–In virology, an infection resulting from the spread of virus from the site of infection to all or most cells of an organism.

Tail–In nematodes: The portion of the body between the anus and the posterior terminus.

Taxon–A taxonomic group of any rank.

Telamon–In nematodes: Rigid, sclerotized portion of the cloacal wall that apparently guides the splcules from the splcular pouch into the cloaca.

Teleutosorus–An old term for telium.

Teleutospore–Search in teliospore.

Teliospore–A usually thick-walled, usually resting spore, which is the site of karyogamy and produces the basidium (in the Uredinales and Ustilaginales).

Telium–The final sorus (stage III) produced in the life cycle of the Uredinales and producing teliospores. (pl. **telia.**)

Teleomorph–The stage characterized by the production of asci/ascospores, basidia/basidiospores, teliospores, or other basidium-bearing organs.

Tessellate–In nematodes: Checkered; a type of cuticular pattern in which the longitudinal ridges are broken by transverse striations into rows of squares.

Testis–A male reproductive organ in which spermatozoa are produced.

Texture–The arrangement of the components of the different tissues, as compact, loose, etc.

Thallic–One of two basic kinds of conidiogenesis; the conidial initial enlarges *after* it has been delimited by one or more septa. The conidium is differentiated from a *whole* cell. (Search in blastic.)

Thallospore–A conidium that has no conidiophore or is not separate from the hypha or conidiophore that produced it.

Thallus–Any simple vegetative plant body that lacks true roots, stems and leaves. (pl. **thalli**.)

Therapy–Principle of plant disease control marked by the cure of disease, as with heat or systemic chemicals.

Threadworm–Nematode.

Tissue–A group of cells of similar structure which performs a special function.

Titillae–In nematodes: Small projections on either side of the distal end of the gubernaculum.

Toadstool–Mushroom.

Tolerance–1. The ability of a plant to sustain the effects of a disease without dying or suffering serious injury or crop loss.

2. The amount of toxic residue allowable in or on edible plant parts under the law.

Tombusvirus–(Siglum of tomato bushy stunt virus). Member of a group of plant viruses with isometric (icosahedral) particles containing one molecule of linear RNA, readily sap-transmitted and also transmitted through soil.

Tospovirus–(Siglum of tomato spotted wilt virus). Member of a small group of plant viruses with spherical particles containing three RNA species, transmitted in nature by thrips in a persistent manner.

Topotype–A specimen collected at the type locality.

Toxicity–The capacity of a compound to produce injury or death.

Toxin–A compound produced by a microorganism and being toxic to a plant or animal.

Trama–The sterile tissue of a basidiocarp.

Transduction–The virus-mediated transfer of host DNA (chromosomal or plasmid) from one host cell (the donor) to another (the recipient). Transduction was first observed in bacteriophage/bacterium systems, but has since also been found to be mediated by certain viruses infecting eukaryotic cells.

Transfection–The successful virus-infection of cells following their inoculation with viral nucleic acid.

Transformant–A cell or organism that has undergone genetic transformation.

Transformation–A process in which exogenous DNA is taken up by a (reciplent) cell or protoplast, in which it may be incorporated into the chromosome (or, *e.g.*, into a plasmid) by homologous recombination or converted into an autonomous replicon. The DNA (transforming or donor DNA) may be a fragment of chromosomal DNA from a related strain, a plasmid, or a viral genome.

Translocation–Transfer of nutrients or virus through the plant.

Transmission–The transfer of a pathogen from one plant to another or from one plant organ to another.

Transpiration–It is the loss of water vapor from the surface of leaves and other aboveground parts of plants.

Transposon–A discrete plece of DNA that can insert itself into other DNA sequences within the cell. The ends of the transposon DNA are usually inverted repeats.

Tretic–Of conidiogenesis: The sort of blastic conidiogenesis in which each conidium is delimited by an extension of the inner wall of the conidiogenous cell. Tretoconidia are solitary or in acropetal chains.

Trichogyne–In some algae, lichens, and fungi, a projection from the female sex organ that receives the male gamete or nuclei before fertilization (karyogamy).

Trifurcate–Bearing three branches or forks.

Triradiate–Having three radiating arms or branches.

Truncate–Having the end squared off or even.

Tumefaction–A plant tumor or gall.

Tumor–An uncontrolled growth of tissue or tissues.

Tylosis–A balloon-like outgrowth from a xylem parenchyma cell that expands into and blocks the lumen of a xylem vessel or a tracheid. (pl. **tyloses**.)

Tymovirus–(Siglum of turnip yellow mosaic virus). Member of a group of plant viruses with isometric (icosahedral) particles containing a single molecule of linear RNA, easily transmitted mechanically and transmitted in nature by beetles.

Type-An object that serves as the basis for the name of a taxon.

Umbilicate–Of the plleus: Having a central navel-like depression, sunken in the center, somewhat funnel-shaped.

Umbo–Of the plleus: A raised, conical to convex knob or mound on the center.

Umbonate–Of the plleus: provided with an umbo or boss.

Uninucleate–Having one nucleus per cell.

Unipolar–At one end only.

Uniseriate–In one row.

Uniseptate–Having one septum.

Unitunicate–Having a single-layered ascus wall.

Urceolate–Pitcher-like, hollow and contracted at the mouth like an urn.

Uredinales–The rust fungi; obligate plant parasites in the Basidiomycotina.

Urediniospore–A binucleate repeating spore borne in the uredinium. (Uredinales) (Also **uredospore, urediospore.**)

Uredinium–The sorus that bears urediniospores in the Uredinales (stage II). (pl. **uredinia.**)

Urediospore–Search in urediniospore.

Uredium–Search in uredinium. (pl. ` **uredia.**)

Uredosorus–Search in uredinium. (pl. **uredosori.**)

Uredospore–Search in urediniospore.

Ustilaginales–The smut fungi; plant parasites in the Basidiomycotina.

Vagina–In nematodes: A canal, lined with cuticle, that connects the uterus or uteri with the female gonospore.

Valid–Of taxonomic names and eplthets: Published in accordance with several articles of the Code of Nomenclature; such names may be legitimate or illegitimate.

Valve–In nematodes: A structure that regulates the rate and/or direction of intake of materials (*e.g.*, the esophago-intestinal valve or cardia).

Valvulated–Having small valves.

var.–Variety.

Variety–In fungi: A rank below subspecies. In bacteria: Formerly a rank equivalent to subspecies; currently an infrasubspecific rank which has no official standing in nomenclature. **vas deferens.** In nematodes: A slender, tube-like gonoduct in the male; unites posteriorly with the rectum to form the cloaca.

Vector–An living agent that transmits a pathogen from an infected plant to an uninfected one.

Vegetative–A cell or structure that is not producing reproductive structures, usually in the assimilative state.

Vegetative reproduction–Asexual reproduction.

Vein banding–A symptom of virus-infected leaves in which tissues along the veins are darker green than other laminar tissue.

Vein clearing–A symptom of virus-infected leaves in which veinal tissue is lighter green than that of healthy plants.

Ventral–Front, or lower surface.

Vermicular–Worm-shaped, thickened and bent in places. (Also vermiculate.)

Vermiculate–Search in vermicular.

Vermiform–Worm-like.

Verrucose–Having small, rounded processes or "warts".

Vertical resistance–In a given cultivar: the existence of differential levels of resistance to different races of a given pathogen.

Vesicle–A bladder-like sac or an evanescent bubble within which zoospores mature; any bubble-like cell or bubble-like membranous structure within a cell.

Vessel–A xylem element or series of such elements whose function is to conduct water and mineral nutrients.

Vestigial–Pertaining to structures or organs that were well developed in an organism's ancestors but have become rudimentary during the course of evolution.

Viable–Living, able to germinate or grow.

Virescence–Greening of tissue that is normally devoid of chlorophyll; the abnormal development of flowers in which all organs are green and partly or wholly transformed into structures like small leaves.

Virescent–A normally white or coloured tissue that develops chloroplasts and becomes green.

Virion–The infectious unit of a virus.

Viroid–Any of numerous kinds of small particles (250-400 nucleotides) of circular, single-stranded RNA that is unencapsidated and encodes no known proteins.

Virulence–The degree of pathogenicity of a given pathogen.

Virulent–Capable of causing a severe disease; strongly pathogenic.

Viruliferous–Used to describe a vector containing a virus and capable of transmitting it.

Virus–Infectious units comprising either RNA or DNA enclosed in a protective coat.

Viviparous–Bearing living young.

Volva–A cup-like structure at the base of a basidiocarp (in the Basidiomycetes).

Volvate–Having a volva.

Vulva–In nematodes: Exterior opening of the mature female's reproductive system (female gonopore); generally appears as a transverse slit on the ventral portion of the nematode.

Western blot–A protein blot. (This term originated as lab jargon; not a generally accepted term.)

Wild type–The phenotype characteristic of the majority of individuals of a species under natural conditions.

Wilt–A disease (or symptom) characterized by a loss of turgidity in a plant (*e.g.*, vascular wilt).

Wilting–Of plant disease: A symptom characterized by loss of turgor, which results in drooplng of leaves, stems, and flowers.

Witches' broom–An abnormal form of plant growth, most common in woody plants, in which there is a profuse outgrowth of lateral buds to give a "witches' broom" appearance. The shoots may be thickened and may bear abnormal leaves.

Woronin body–A rounded organelle occurring near septa in at least some ascomycetes or deuteromycetes.

Yeast–A unicellular member of the Endomycetales; sometimes used for a non-motile unicellular stage (*e.g.*, of dimorphic animal parasites).

Yellowing–A symptom characterized by the turning yellow of plant tissues that were once green.

Yellows–Any of a wide variety of plant diseases in which a major symptom is a uniform or non-uniform yellowing of leaves and/or other plant components. Yellows may be caused by fungi (*e.g.*, celery yellows), viruses (*e.g.*, sugar beet yellows virus), bacteria (*e.g.*, coconut lethal yellowing), protozoa (*e.g.*, hartrot), splroplasmas or phytoplasmas.

Zoosporangium–A sporangium within which zoospores are produced. (pl. **zoosporangia**.)

Zoospore–An asexual, motile spore that bears one or two flagella.

Zygomycetes–Search in Zygomycotina.

Zygomycotina–A subdivision of fungi characterized by the formation of a thick-walled resting spore formed after the fusion of two equal gametangia.

Zygosporangium–In the Zygomycotina: A usually thick-walled, often ornamented, multinucleate resting sporangium formed following anastomosis of gametangia arising from compatible mycelia (in heterothallic species) or from the same mycelium (in homothallic species)..

Zygospore–A sexual resting spore formed within a zygosporangium.

Zygote–A diploid cell resulting from the fusion of two gametes; a diploid nucleus resulting from the union of two haploid nuclei.

Chapter 7
Plant Breeding & Genetics Terms

Chapter 10
Water and Water-Related Terms

Absorbed water–Water held mechanically in a soil mass and having physical properties not substantially different from ordinary water at the same temperature and pressure.

Acid mine drainage–The oxidation of sulphides to sulphuric acid often causes low pH drainage water from certain mines. Mine drainage can also contain high concentration of metal ions.

Acid rain–Acid rain is essentially a rainfall with a pH value of less than 7.0. One of the sources of acid rain is the combination of rain and sulphur dioxide emissions, which are a by-product of combustion of fossil fuels. It is also referred to as acid deposition and wet deposition.

Active absorption–Absorption of water and other materials involving an expenditure of cellular energy in contrast to imbibition and diffusion.

Adhesion–It is the molecular attraction which holds the surfaces of two substances in contact,eg., water and sand particle.

Algae–These are simple rootless plants that grow in sunlit waters in relative proportion to the amounts of nutrients available. They can affect water quality adversely by lowering the dissolved oxygen in the water. They are food for fish and small aquatic animals.

Algae blooms–It refers to the rapid growth of algae on the surface of lakes, streams, or ponds; stimulated by nutrient enrichment.

Alkali–Any strongly basic substance of hydroxide and carbonate that is soluble in water and increases the pH of a solution is called alkali. Examples–soda, potash, etc.

Anhydrous–Being without water, especially water of crystallization.

Aquatic ecosystem–It is a basic ecological unit composed of living and nonliving elements interacting in an aqueous milieu.

Aquifer–The underground layer of water-soaked sand and rock that acts as a water source for a well is called aquifer. It is described as artesian (confined) or water table (unconfined).

Arid–The regions where precipitation is insufficient in quantity for most crops and where agriculture is impractical without irrigation are arid regions.

Atmosphere–The layer of gases surrounding the earth composed of considerable amounts of nitrogen, hydrogen, and oxygen, is atmosphere.

Atmospheric water–Water present in the atmosphere either as a solid (snow, hail), liquid (rain) or gas (fog, mist) is atmospheric water.

Bioaccumulation (bioconcentration)–It is a term used to describe a process that occurs when levels of toxic substances increase in an organism over time, due to continued exposure.

Biodegradable–Substances that are capable of being broken down by living organisms into inorganic compounds.

Biological diversity (biodiversity)–These are the variety of different species, the genetic variability of each species, and the variety of different ecosystems that they form.

Biomagnification (biological magnification)–A cumulative increase in the concentrations of a persistent substance in successively higher levels of the food chain is known as biomagnification.

Biota–Biota is the sum total of the plants, microorganisms, and animals of a certain area or region.

Bog–A type of wetland that accumulates appreciable peat deposits is bog. It depends primarily on precipitation for its water source and is usually acidic and rich in plant matter, with a conspicuous mat or living green moss.

Booster pump–Pump installed on a pipeline to raise the pressure of water on the discharge side of the pump.

Boundary water–A river or lake that is part of the boundary between two or more countries or provinces that have rights to the water.

Capillary potential–It is the required to pull a unit mass of water away from a unit mass of soil.

Capillary water–The water retained in the fine soil capillaries after seepage and percolation processes are over is known as capillary water.

Capillary movement of water–Upward movement of water subjected to evapo-transpiration in soil capillaries is called capillary movement of water. Finer the soil capillaries, greater is the rise in water column and vice versa.

Check irrigation–Method of irrigation where the field or the orchard to be irrigated is divided into a series of checks by earth ridges and the water flows from one check to another along the slope.

Climate–Meteorological elements that characterize the average and extreme conditions of the atmosphere over a long period of time at any one place or region of the earth's surface is called climate.

Climate change–It is the slow variations of climatic characteristics over time at a given place.

Coliform bacteria–A group of bacteria used as an indicator of sanitary quality in water. Exposure to these organisms in drinking water causes diseases like cholera and other water borne diseases.

Combined sewers–It is a sewer that carries both sewage and storm water runoff.

Condensation–It is the process by which a vapour becomes a liquid or solid. The process is just opposite to evaporation. In meteorological usage, this term is applied only to the transformation from vapour to liquid.

Conservation–Conservation is the continuous protection and management of natural resources in accordance with principles that assure their optimum long-term economic and social benefits.

Consumptive use–1. It is the amount of water required by plants to meet their need for water in transpiration and growth. It is also referred as evapotranspiration.

2. The difference between the total quantity of water withdrawn from a source for any use and the quantity of water returned to the source; *e.g.*, the release of water into the atmosphere; the consumption of water by humans, animals, and plants; and the incorporation of water into the products of industrial or food processing.

Contaminant–Contaminant is any physical, chemical, biological, or radiological substance or matter that has an adverse affect on air, water, or soil.

Contour irrigation–Method of irrigation where the water is applied to a field or orchard which has been divided into a series of strips, by ridges located along contours or lines of equal elevation.

Cooling tower–A structure that helps remove heat from water used as a coolant; *e.g.*, in electric power generating plants.

Critical period–The period during which a crop is affected severely due to moisture stress and the loss can not be compensated by adequate moisture supply.

Cubic metre per second (m^3/s)–It is a unit expressing rate of discharge, typically used in measuring streamflow. One cubic metre per second is equal to the discharge in a stream of a cross section one metre wide and one metre deep, flowing with an average velocity of one metre per second.

Dam–Dam is a structure of earth, rock, concrete, or other materials designed to retain water, creating a pond, lake, or reservoir.

Dead water–Standing or still water.

Delta–It is a fan-shaped alluvial deposit at a river mouth formed by the deposition of successive layers of sediment.

Demand–The numerical expression of the desire for goods and services associated with an economic standard for acquiring them.

Depletion–Loss of water from surface water reservoirs or groundwater aquifers at a rate greater than that of recharge.

Dew–It is the result of condensation of a part of vapour into a liquid form on coming in contact with cool surface.

Diara land–It is an area which is located in river basins that get submerged during rainy season and the soil is used for crop production only during rabi and/or sometimes during zaid seasons. It is also called as Kachhar land.

Diffusion–The upward or downward movement of soluble salts in a profile,without movement of water, is known as diffusion.

Dioxin–Any compounds within a family known chemically as dibenzo-p-dioxins. The potential toxicity and contamination in commercial products arises serious concern about them.

Discharge–In the simplest form, discharge means outflow of water. The use of this term is not restricted as to course or location, and it can be used to describe the flow of water from a pipe or from a drainage basin. Other words related to it are runoff, streamflow, and yield.

Dissolved oxygen (DO)–This is the amount of oxygen freely available in water and necessary for aquatic life and the oxidation of organic materials. It is a indicator of pollution measurement.

Dissolved solids (DS)–Very small pieces of organic and inorganic material contained in water are called as dissolved solids. Excessive amounts make water unfit to drink or limit its use in industrial processes.

Diversion–The transfer of water from a stream, lake, aquifer, or other source of water by a canal, pipe, well, or other conduit to another watercourse or to the land, as in the case of an irrigation system.

Domestic use–The quantity of water used for household purposes such as washing, food preparation, and bathing.

Drainage–Removal of excess surface or groundwater from land by means of surface or surface or sub-surface drain.

Drainage basin–*Search* Watershed.

Dredge rate–The material excavated from lake, river, or channel bottoms during dredging.

Dredging–The removal of material from the bottom of water bodies using a scooping machine. This disturbs the ecosystem and causes silting that can kill aquatic life.

Drought –1. A continuous and lengthy period during which no significant precipitation is recorded.

2. *In plants*–It is a condition of insufficient moisture supply under which they fail to develop and mature properly. It may be caused by soil, atmosphere and/or both.

Dry deposition–Emissions of sulphur and nitrogen oxides that, in the absence of water in the atmosphere (*i.e.*, rain), settle to the ground as particulate matter.

Dry farming–The practice of raising successful and profitable rainfed crops in areas which receive an annual rainfall of 50 cm or less is known as dry farming.It is also called as dryland agriculture.

Duty of water–It is the area that may be irrigated by a definite quantity of water throughout the life span of the crop.

Dyke–It is an artificial embankment constructed to prevent flooding.

Ecosystem–A system formed by the interaction of a group of organisms and their environment.

Effective rainfall–Precipitation falling during the growing period of the crop that is available to meet he evapotranspiration requirements of crop. It does not include the percolation loss below the root zone or that through surface runoff.

Effluent–The sewage or industrial liquid waste that is released into natural water by sewage treatment plants, industry, or septic tanks.

Environment–The sum total of the external factors, conditions, and influences that affect an organism or a community.

Environmental assessment–The critical appraisal of the likely effects of a proposed project, activity, or policy on the environment, both positive and negative. It is important for controlling pollution.

Environmental monitoring–The process of checking, observing, or keeping track of something for a specified period of time or at specified intervals.

Erosion–Detachment and transportation of soil particles from their original place to any other place through water, wind, ice, or any other geological agent is called as erosion.

Estuary–Regions of interaction between rivers and nearshore ocean waters, where tidal action and river flow create a mixing of fresh water and saltwater. These areas may include bays, mouths of rivers, salt marshes, and lagoons. These brackish water ecosystems shelter and feed marine life, birds, and wildlife.'

Eutrophic lake–Shallow, murky bodies of water that have excessive concentrations of plant nutrients causing excessive algal production.

Eutrophication–The natural process by which lakes and ponds become enriched with dissolved nutrients, resulting in increased growth of algae and other microscopic plants.

Evaporation–It is the loss of water in vapour form either from soil or water surface subjected due to high atmospheric temperature associated with lower relative humidity. The evaporational loss of water is measured by open pan evaporimeter.

Evapotranspiration–It is an additive or cumulative volume/quantity of water lost due to both evaporation and transpiration from the soil surface. When it becomes difficult to separate out both the losses from one another then an additive value is expressed and the term evapotranspiration is used.

Fen–A type of wetland that accumulates peat deposits. Fens are less acidic than bogs, deriving most of their water from groundwater rich in calcium and magnesium.

Field capacity–The percentage of water remaining in soil two or three days after having been saturated and after free drainage has practically ceased.

Flood–The temporary inundation of normally dry land areas resulting from the overflowing of the natural or artificial confines of a river or other body of water.

Flood damage–The economic loss caused by floods, including damage by inundation, erosion, and/or sediment deposition. Damages also include emergency costs and business or financial losses. Evaluation may be based on the cost of replacing, repairing, or rehabilitating; the comparative change in market or sales value; or the change in the income or production caused by flooding.

Flood forecasting–Prediction of stage, discharge, time of occurrence, and duration of a flood, especially of peak discharge at a specified point on a stream, resulting from precipitation and/or snowmelt.

Flood fringe–The portion of the floodplain where water depths are shallow and velocities are low is known as flood fringe.

Flood peak–The highest magnitude of the stage of discharge attained by a flood. It is also called peak stage or peak discharge.

Floodplain–Floodplain is any normally dry land area that is susceptible to being inundated by water from any natural source. This area is usually low land adjacent to a stream or lake.

Floodproofing–Any combination of structural and nonstructural additions, changes, or adjustments to structures that reduce or eliminate flood damage.

Floodway–The channel of a river or stream and those parts of the adjacent floodplain adjoining the channel that are required to carry and discharge the base flood.

Flow–The rate of water discharged from a source; expressed in volume with respect to time, *e.g.*, m^3/s.

Flow augmentation–The addition of water to a stream, especially to meet instream flow needs.

Food chain–A sequence of organisms, each of which uses the next, lower member of the sequence as a food source (from producers to consumers to decomposers).

Food web–The complex intermeshing of individual food chains in an ecosystem.

Free water–Amount of water removed in drying a solid to its equilibrium water content.

Fresh water–Water that generally contains less than 1000 milligrams per litre of dissolved solids such as salts, metals, nutrients, etc.

Glacial drift–Rock debris that has been transported by glaciers and subsequently deposited, either from ice or from melting water.

Glacier–A huge mass of ice, formed on land by the compaction and re-crystallization of snow, which moves very slowly downslope or outward due to its own weight.

Gley soil–Soil developed under conditions of poor drainage resulting into grey or mottle colour due to reduction in iron or other elements.

Greenhouse effect–The warming of the earth's atmosphere caused by a build-up of carbon dioxide or other trace gases is known as green house effect. Many scientists believe that this build-up allows light from the sun's rays to heat the earth but prevents a counterbalancing loss of heat.

Groundwater–The supply of fresh water found beneath the earth's surface (usually in aquifers) that is often used for supplying wells and springs.

Groundwater recharge–The inflow to an aquifer is called as groundwater recharge.

Guttation–Exudation of liquid water from plant leaves.

Habitat–The native environment where a plant or animal naturally grows or lives.

Hard water–Water which contains certain minerals like calcium and magnesium sulphates, chlorides, carbonates, in such quantities as to precipitate rather than lather when soap is added.

Hazardous waste–Waste that poses a risk to human health or the environment and requires special disposal techniques to make it harmless or less dangerous.

Heavy water–A special water (D_2O) in which the hydrogen of the water molecule consists entirely of the heavy-hydrogen isotope of mass 2 (deuterium). It is used as moderator in certain types of nuclear reactor.

Hydroelectricity–Electric energy produced by water-powered turbine generators.

Hydrograph–Graph on which stream flow is plotted as a function of time.

Hydrologic cycle–The constant circulation of water from the sea, through the atmosphere, to the land, and back to the sea by over-land, underground, and atmospheric routes.

Hydrology–The science of waters of the earth; water's properties, circulation, principles, and its distribution.

Hydroponics–Cultivation of plants in nutrient solutions rather than a soil.

Hygroscopic water–It is a very thin film of water of about 4 or 5 millimicrons at the solid liquid interfaces on the soil particles or granules which is in a non-liquid state and as such is immobile.

Hygroscopic coefficient–The amount of moisture in dry soil when it is in equilibrium with some relative humidity near a saturated atmosphere(about 98 per cent), expressed in terms of percentage on the basis of oven dry soil.

Imbibitional water–Water held within the lattice of the colloidal matter of the soil.

Indicator plant–A plant which indicates acute moisture stress in the soil is known as moisture indicator plant.

Infiltration–The movement of water into soil or porous rock. Infiltration occurs as water flows through the larger pores of rock or between soil particles under the influence of gravity, or as a gradual wetting of small particles by capillary action.

Inflow–The entry of extraneous rainwater into a sewer system from sources other than infiltration, such as basement drains, sewer holes, storm drains, and street washing.

Instream use–Uses of water within the stream channel, *e.g.*, fish and other aquatic life, recreation, navigation, and hydroelectric power production.

Integrated resource planning–It is the management of two or more resources in the same general area. Commonly it includes water, soil, timber, grazing land, fish, wildlife, and recreation.

Interbasin transfer–The diversion of water from one drainage basin to one or more other drainage basins.

Irrigation–The controlled application of water to cropland, hayland, and/or pasture to meet the water requirement of the crop is called as irrigation.

Irrigation efficiency–It is the ratio of water actually consumed by crops on an irrigated area to the amount of water diverted from the source on to the area.

$$\text{Irrigation efficiency} = \frac{\text{Consumed water by crops}}{\text{Total water applied to the field}}$$

Irrigation requirement–In addition to rainfall and soil profile reserves, the amount of water supplied to the land surface for meeting the water need of crops for optimum growth.

Jökulhlaup–Destructive flood that occurs as the result of the rapid ablation of ice by volcanic activity beneath the ice of a large glacier.

Kilowatt (kW)–A unit of electrical power equal to 1000 watts or 1.341 horsepower (H.P.).

Kilowatt hour (kWh)–One kilowatt of power applied for one hour.

Lagoon–(1) A shallow pond where sunlight, bacterial action, and oxygen work to purify wastewater. (2) A shallow body of water, often separated from the sea by coral reefs or sandbars.

Lake–Any inland body of standing water, usually fresh water, larger than a pool or pond; a body of water filling a depression in the earth's surface.

Leaching–The removal of soluble organic and inorganic substances from the topsoil downward by the action of percolating water.

Leaching requirement–Fraction of water (irrigation) entering the soil that must pass through the root zone in order to prevent salinity from exceeding a specified value.

Life saving irrigation–Irrigation that saves the plants from death under drought condition.

Litre–The basic unit of measurement for volume in the metric system; equal to 61.025 cubic inches or 1.0567 liquid quarts and its sign is L.

Lysimeter–A device for measuring percolation and leaching losses from a column of soil under controlled conditions.

Marsh–A type of wetland that does not accumulate appreciable peat deposits and is dominated by herbaceous vegetation. Marshes may be either fresh water or saltwater and tidal or non-tidal.

Maximum available water–Quantity of water represented as the difference between field capacity and writing coefficient. This can be readily extracted by the plant and used for normal growth.

Maximum retention capacity–It is the moisture percentage in soil when its total pore-space is completely filled with water.

Megawatt–A unit of electricity equivalent to 1000 kilowatts (KW).

Model–A simulation, by descriptive, statistical, or other means, of a process or project that is difficult or impossible to observe directly.

Moisture tension–It is the equivalent negative pressure in the soil water.

Movement of soil water–Water movement from surface to water table and back to the atmosphere through the soil occurring due to capillary adjustment, percolation, vapour equalizations etc, is known as movement of soil water.

NAPLs–Nonaqueous phase liquids; *i.e.*, chemical solvents such as trichloroethylene (TCE) or carbon tetrachloride –often toxic. Many of the most problematic NAPLs are DNAPLs –dense nonaqueous phase liquids.

Natural flow–The flow of a stream as it would be if unaltered by upstream diversion, storage, import, export, or change in upstream consumptive use caused by development.

Navigable waters–Traditionally, waters sufficiently deep and wide for navigation by all, or specific sizes of, vessels.

Oligotrophic lake–Deep, clear lakes with low nutrient supplies. They contain little organic matter and have a high dissolved oxygen level.

Osmotic–It is a type of pressure exerted in living bodies as a result of unequal concentration of salt on both sides of cell wall or any membrane. The movement of water occurs from the area having the least concentration through the membrane into the area having the highest concentration.

Outlet–Downstream opening of discharge end of a pipe, culvert, sewer or canal.

Parts per million (PPM)–The number of "parts" by weight of a substance per million parts of water. This unit is commonly used to represent pollutant concentrations. Large concentrations are expressed in percentages.

Pathogenic microorganisms–Microorganisms that can cause disease in other organisms or in humans, animals, and plants.

Percolation–The downward movement of water through the subsurface to the zone of saturation due to gravitational pull. It also known as infiltration or leaching of water.

Permafrost–Perennially frozen layer in the soil, found in alpine, arctic, and antarctic regions.

Permeability–Rate at which water moves through a soil under a standard pressure gradient. It is measured by the amount of water traversing 1 square cm of soil surface under a pressure gradient of 1 dyne per cm.

Pesticide–A substance or mixture of substances intended for preventing, destroying, repelling, or mitigating any pest. Also, any substance or mixture of substances intended to regulate plant or leaf growth. Pesticides can accumulate in the food chain and/or contaminate the environment if misused.

pF–The logarithm of the soil moisture tension expressed in centimeters height of a water column.

pH–An expression of both acidity and alkalinity on a scale of 0 to 14, with 7 representing neutrality; numbers less than 7 indicate increasing acidity and numbers greater than 7 indicate increasing alkalinity.

Photosynthesis–The manufacture by plants of carbohydrates and oxygen from carbon dioxide and water in the presence of chlorophyll, using sunlight as an energy source.

Phytoplankton–Usually microscopic, free floating aquatic plants, sometimes consisting of only one cell.

Plankton–Tiny plants and animals that live in water.

Polychlorinated biphenyls (PCBs)–A group of chemicals found in industrial wastes.

Pond–A small natural body of standing fresh water filling a surface depression, usually smaller than a lake.

Precipitation–Water falling, in a liquid or solid state, from the atmosphere to a land or water surface.

Rain–Water falling to earth in drops that have been condensed from moisture in the atmosphere.

Rain gauge–A device for measuring the depth of rainfall.

Rainfed farming–It is a type of farming which entirely depends upon rain water and a profitable production is achieved. Unlike dry farming (where the upper limit of annual precipitation is fixed to be 50 cm), there is no upper limit of total annual rainfall.

Receiving waters–A river, ocean, stream, or other watercourse into which wastewater or treated effluent is discharged.

Recharge–The processes involved in the addition of water to the zone of saturation; also the amount of water added.

Reservoir–A pond, lake, or basin (natural or artificial) that stores, regulates, or controls water.

River–River is a natural stream of water of substantial volume.

River basin–A term used to designate the area drained by a river and its tributaries.

Runoff–The amount of precipitation appearing in surface streams, rivers, and lakes; defined as the depth to which a drainage area would be covered if all of the runoff for a given period of time were uniformly distributed over it.

Saltwater intrusion–The invasion of fresh surface water or groundwater by saltwater.

Sanitary sewers–Underground pipes that carry off only domestic or industrial waste but not storm water.

Sediment–Fragmented organic or inorganic material derived from the weathering of soil, alluvial, and rock materials; removed by erosion and transported by water, wind, ice, and gravity.

Sedimentation–The deposition of sediment from a state of suspension in water or air.

Seiche–A periodic oscillation, or standing wave, in an enclosed water body the physical dimensions of which determine how frequently the water level changes.

Septic tank–Tank used to hold domestic wastes when a sewer line is not available to carry them to a treatment plant; part of a rural on-site sewage treatment system.

Sewage–The waste and wastewater produced by residential and commercial establishments and discharged into sewers.

Sewage system–Pipelines or conduits, pumping stations, force mains, and all other structures, devices, and facilities used for collecting or conducting wastes to a point for treatment or disposal.

Sewer–A channel or conduit that carries wastewater and storm water runoff from the source to a treatment plant or receiving stream.

Sewerage–The entire system of sewage collection, treatment, and disposal.

Silt–Fine particles of sand or rock that can be picked up by the air or water and deposited as sediment.

Sludge–A semi-solid residue from any of a number of air or water treatment processes.

Solvent–Substances (usually liquid) capable of dissolving or dispersing one or more other substances.

Spillways–Conduit in or around a dam for the escape of excess water.

Spoils–Dirt or rock that has been removed from its original location, destroying the composition of the soil in the process, as with strip-mining or dredging.

Spring–An area where groundwater flows naturally onto the land surface.

Storm sewer–A system of pipes (separate from sanitary sewers) that carry only water runoff from building and land surfaces.

Stream–Any body of running water moving under gravity flow through clearly defined natural channels to progressively lower levels.

Stream line flow–When as much of the liquid leaves a pipe as enters it per second, the velocity at every point in the liquid remains constant and the liquid is said to have a streamline-flow.

Supplemental irrigation–Watering of crops in regions where normal rainfall ordinarily supplies most of the moisture. It is normally given in dry periods to prevent retardation of growth.

Surface water–All water naturally open to the atmosphere (rivers, lakes, reservoirs, streams, impoundments, seas, estuaries, etc.); also refers to springs, wells, or other collectors that are directly influenced by surface water.

Surface tension–The attractive force of water molecules for each other manifested at all liquid-air interfaces is known as surface tension.

Suspended solids (SS)–Defined in waste management, these are small particles of solid pollutants that resist separation by conventional methods. Suspended solids (along with biological oxygen demand) are a measurement of water quality and an indicator of treatment plant efficiency.

Swamp–A type of wetland that is dominated by woody vegetation and does not accumulate appreciable peat deposits. Swamps may be fresh water or saltwater and tidal or nontidal.

Temperature–The degree of hotness or coldness.

Thermal pollution–The impairment of water quality through temperature increase; usually occurs as a result of industrial cooling water discharges.

Toxic–Harmful to living organisms.

Transpiration–It is a physiological process by which water absorbed by plants, usually through the roots, is evaporated into the atmosphere from the plant surface, principally from the leaves, just to protect them from rising atmospheric temperature.

Tributary–A stream that contributes its water to another stream or body of water.

Tsunami–A Japanese term that has been adopted to describe a large seismically generated sea wave capable of considerable destruction in certain coastal areas, especially where sub-marine earthquakes occur.

Turbidity–Cloudiness caused by the presence of suspended solids in water; an indicator of water quality.

Underground storage tank–A tank located all or partially underground that is designed to hold gasoline or other petroleum products or chemical solutions.

Urban runoff–Storm water from city streets and adjacent domestic or commercial properties that may carry pollutants of various kinds into the sewer systems and/or receiving waters.

Vapour–The gaseous phase of substances that are liquid or solid at atmospheric temperature and pressure, *e.g.*, steam.

Vapour pressure–Partial pressure exerted by water vapour in the air and is independent of the other gases.

Waste disposal system–A system for the disposing of wastes, either by surface or underground methods; includes sewer systems, treatment works, and disposal wells.

Wastewater–Water that carries wastes from domestic uses, businesses, and industries; a mixture of water and dissolved or suspended solids.

Wastewater treatment plant–A facility containing a series of tanks, screens, filters, and other processes by which pollutants are removed from water.

Water (H_2O)–An odourless, tasteless, colourless liquid formed by a combination of hydrogen and oxygen; forms streams, lakes, and seas, and is a major constituent of all living matter.

Water conservation–It is the care, preservation, protection, and wise use of water.

Water contamination–Impairment of water quality to a degree that reduces the usability of the water for ordinary purposes or creates a hazard to public health through poisoning or the spread of diseases.

Water harvesting–It is the collection and preservation of rain water for the purpose of profitable production of crops particularly in the dry lands.

Water holding capacity–Weight of water held by a given quantity of absolutely dry soil when saturated.

Water logging–Soil saturated with free water which may get accumulated on the ground surface.

Water management–The study, planning, monitoring, and application of quantitative and qualitative control and development techniques for long-term, multiple use of the diverse forms of water resources.

Water pollution–Industrial and institutional wastes and other harmful or objectionable material in sufficient quantities to result in a measurable degradation of the water quality.

Water potential–Difference between the activity of water molecules in pure distilled water at normal atmospheric pressure at 30°C (standard conditions) and the activity of water molecules in any other system; the activity of these water molecules may be greater (positive) or less (negative) compared to that under standard conditions.

Water quality–A term used to describe the chemical, physical, and biological characteristics of water with respect to its suitability for a particular use.

Water quality guidelines–Specific levels of water quality that, if reached, are expected to render a body of water suitable for its designated use. The criteria are based on specific levels of pollutants that would make the water harmful if used for drinking, swimming, farming, fish production, or industrial processes.

Water requirement of crops–It is the volume of water which is required to produce a crop in a specific period of time. It includes evapotranspirational and other water losses due to percolation, seepage, runoff and water required for seed bed preparation.

Water supply system–The collection, treatment, storage, and distribution of potable water from source to consumer.

Water table–The top of the zone of saturation.

Watershed–The land area that drains into a stream.

Weir–Dam across a stream for diverting or for measuring the flow.

Weir notch–Opening in a weir for the passage of water.

Well–A pit, hole, or shaft sunk into the earth to tap an underground source of water.

Wet deposition–See *acid rain*.

Wetlands–Lands where water saturation is the dominant factor determining the nature of soil development and the types of plant and animal communities living in the surrounding environment. Other common names for wetlands are bogs, ponds, estuaries, and marshes.

Wilting point–It is that moisture percent in the soil through which the plants can hardly meet their transpirational requirement of water. In plants two types of wilting are there:

1. Temporary wilting–When the plants start showing moisture scarcity by appearance of leaf rolling during the hotter part of the day but recover during evening.

2. Permanent wilting–This stage comes after the plants have crossed the temporary wilting and if the water is not supplied to the plants the young leaves and tender parts get dried or burnt which do not recover even after irrigating them.

Withdrawal use–The act of removing water from surface water or groundwater sources in order to use it.

Xerophytes–These are drought resistant plants which grows in dry habitat or on extremely dry soils or soil materials.

Zooplankton–Tiny aquatic animals eaten by fish.

Zone of saturation–A subsurface zone in which all the pores or the material are filled with groundwater under pressure greater than atmospheric pressure.

Chapter 11
Quality Management Terms

Accreditation–Certification by a duly recognized body of the facilities, capability, objectivity, competence, and integrity of an agency, service or operational group or individual to provide the specific service(s) or operation(s) needed.

Accredited Registrars–Qualified organizations certified by a national body to perform audits to the **QS–9000** and to register the audited facility as meeting these requirements for a given commodity.

AIAG–Automotive Industry Action Group

AQL–Acceptable quality level

Assessment–An evaluation process including a document review, an on-site audit and an analysis and report. (Search in **Quality audit**).

Assignable cause–Search in **Special cause**

ASTM–American Society for Testing and Materials

Attributes–Qualitative data that can be counted for recording and analysis. Examples include characteristics such as the presence of a required label and the installation of all required fasteners.

Audit–1. An onsite verification activity used to determine the effective implementation of a supplier's documented quality system.

2. Systematic, independent and documented process for obtaining audit evidence and evaluating it objectively to determine the extent to which audit criteria are fulfilled.

Audit client–Organization or person requesting an audit.

Audit conclusion–Outcome of an audit provided by the audit team after consideration of the audit objectives and all audit findings.

Audit criteria–Set of policies, procedures or requirements used as a reference (while conducting an audit).

Auditee–Organization being audited.

Audit evidence–Records, statements of fact or other information which are relevant to the audit criteria and verifiable.

Audit findings–Results of the evaluation of the collected audit evidence against audit criteria.

Auditor–Person with the competence to conduct an audit is an auditor.

Audit program–Set of one or more audits planned for a specific time frame and directed towards a specific purpose.

Audit team–One or more auditors conducting an audit.

Availability–The ability of an item to perform its designated function when required for use.

Average or mean–The most common expression of the centering of a distribution. It is calculated by totaling the observed values and dividing by the number of observations.

Benchmark Data–The results of an investigation to determine how competitors and/or best-in-class companies achieve their level of performance.

Bimodal Distribution–A distribution with two identifiable curves within it, indicating a mixing of two populations such as different shifts, machines, workers, etc.

BS–British Standard.

BSI–British Standards Institution.

CAI–Computer aided inspection.

Capability–1. The total range of inherent variation in a stable process. (Search in Process Capability)

2. Ability of an organization, system or process to realize a product that will fulfill the requirements for that product.

CASCO–ISO Committee on Conformity Assessments.

CC–Critical characteristic.

CE Mark–European Union product safety certification symbol: CE

CEN–European Committee for Standardization

CENELEC–European Committee for Electro-technical Standardization

Certificate of compliance–A document signed by an authorized party affirming that the **supplier** of a product or service has met the requirements of the relevant specifications, contract, or regulation.

Certificate of conformance (Certificate of conformity)–A document signed by an authorized party affirming that a **product or service** has met the requirements of the relevant specifications, contract, or regulation.

Certification–The procedure and action by a duly authorized body of determining, verifying, and attesting in writing to the qualifications of personnel, processes, procedures, or items in accordance with applicable requirements.

Characteristic–Distinguishing feature

CIM–Computer Integrated Manufacturing

Common Cause–A source of variation that is always present as part of the random variation inherent in the process itself. Its origin can usually be traced to an element of the system which only management can correct.

Competence–Demonstrated ability to apply knowledge skills.

Compliance –1. An affirmative indication or judgment that the **supplier** of a product or service has met the requirements of the relevant specifications, contract, or regulation; also the state of meeting the requirements.

2. Any raw material, substance, piece, part, software, firmware, labeling, or assembly which is intended to be included as part of the finished, packaged, and labeled device.

Concession–Permission to use or release a product that does not conform to specified requirements.

Conformance–An affirmative indication or judgment that a product or service has met the requirements of the relevant specifications, contract, or regulation; also the state of meeting the requirements.

Conformity–The fulfilling by an item or service of specification requirements.

Continual improvement–Recurring activity to increase the ability to fulfill requirements.

Control Chart–A graphic representation of a characteristic of a process, showing plotted values of some statistic gathered from that characteristic, and one or two control limits.

Control Limit–A line (or lines) on a control chart used as a basis for judging the significance of the variation from subgroup to subgroup. Variation beyond a control limit is evidence that special causes are affecting the process. Control limits are calculated from process data and are not to be confused with engineering specifications.

Control Plans–Written descriptions of the systems for controlling parts and processes.

Correction–Action to eliminate a detected nonconformity.

Corrective action–Action to eliminate the cause of a detected nonconformity or other undesirable situation.

Corrective Action Plan–A plan for correcting a process or part quality issue.

CQA–Certified Quality Auditor.

CQE–Certified Quality Engineer.

CQMgr–Certified quality manager.

CRE–Certified Reliability Engineer.

Customer–Organization or person that receives a product.

Customer satisfaction–Customer's perception of the degree to which the customer's requirements have been fulfilled.

Defect–Non-fulfillment of a requirement related to an intended or specified use.

Dependability–1. The state of being counted on or trusted.

2. Collective term used to describe the availability performance and its influencing factors: reliability performance, maintainability performance and maintenance support performance.

Design and development–Set of processes that transforms requirements into specified characteristics or into the specification of a product, process or system.

Design Failure Mode and Effects Analysis (DFMEA)–An analytical technique used by a design responsible engineer/team as a means to assure, to the extent possible, that potential failure modes and, their associated causes/ mechanisms have been considered and addressed.

Design for Manufacturability and Assembly–A simultaneous engineering process designed to optimize the relationship between design function, manufacturability, and ease of assembly.

Design input–The physical and performance requirements of a device that are used as a basis for device design.

Design Review–1. A proactive process to prevent problems and misunderstandings.

2. A formal, documented, comprehensive, and systematic examination of a design to evaluate the design requirements and the capability of the design to meet these requirements and to identify problems and propose solutions.

Design Validation–1. Testing to ensure that products conforms to defined user needs and/or requirements. Design validation follows successful design verification and is normally performed on the **final** product under defined operating conditions. Multiple validations may be performed if there are different intended uses.

2. Establishing by objective evidence that device specifications conform with user needs and intended use(s).

Design Verification–Testing to ensure that all design outputs meet design input requirements. Design verification may include activities such as:

Design review

Performing alternate calculations

Understanding tests and demonstrations

Review of design stage documents before release

Detection or inspection–A past-oriented strategy that attempts to identify unacceptable output after it has been produced and separate it from the good output. (Search in Prevention and Nonconforming)

Deviation permit–1. Written authorization, prior to production or provision of a service, to depart from specified requirements for a specified quantity or for a specified time.

2. Permission to depart from the originally specified requirements of a product prior to realization.

DFA–Design for assembly.

DFM–Design for manufacturing.

DFMEA–Design Failure Mode and Effects Analysis

DIN–Germany Standards Institute.

Distribution–The population (universe) from which observations are drawn, categorized into cells, and form identifiable patterns. It is based on the concept of variation that states that anything measured repeatedly will arrive at different results. These results will fall into statistically predictable patterns. A bell-shaped curve (normal distribution) is an example of a distribution in which the greatest number of observations occurs in the center with fewer and fewer observations falling evenly on either side of the average.

Document–Information and its supporting medium.

Documentation–Written material defining the process to be followed (*e.g.* test procedure, quality manual, operation sheets).

DOE–Design of Experiments.

Durability–The probability that an item will continue to function at customer expectation levels, at the useful life without requiring overhaul or rebuild due to wearout.

EC–European Community

EFTA–European Free Trade Association

EN–European Standard

EQS–European Committee for Quality System Assessment and Certification

Establish–Define, document (in writing or electronically), and implement.

ETSI–European Telecommunications Standards Institute

Efficacy–(search in effectiveness below)

Efficiency–Relationship between the result achieved and the resources used.

Effectiveness–Extent to which planned activities are realized and planned results achieved.

FMEA method (FMECA)–Failure Mode and Effect (and Criticality) Analysis, a powerful method of risk assessment and failure analysis for use in risk management and product liability control.

Frequency distribution–A statistical table that presents a large volume of data in such a way that the central tendency (average/mean/median) and distribution are clearly displayed.

FTA Fault–Tree Analysis

Functional Verification Testing to ensure the part conforms to all customer and supplier engineering performance and material requirements.

Generally implied–Custom or common practice for the organization, its customers and other interested parties, that the need or expectation under consideration is implied.

GD&T–Geometric Dimensioning and Tolerancing

GMP–Good Manufacturing Practice.

GR&R–Gage Repeatability and Reproducibility

Grade–Category or rank given to different quality requirements for products, processes or systems having the same functional use.

Hardware–Tangible, discrete product with distinctive form.

Histogram–Search in Frequency distribution

Information–Meaningful data

Infrastructure (of an organization) System of facilities, equipment and services needed for the operation of an organization.

Inspection–Activities, such as measuring, examining, testing, gaging one or more characteristics of a product or service, and comparing these with specified requirements to determine conformity.

IEC–International Electro-technical Commission

IEEE–Institute of Electrical and Electronic Engineers

Interested party–Person or group having an interest in the performance or success of an organization.

ISO–International Organization for Standards.

ISO 14000–International environmental management system standard administered by ISO

ISO 9000–International Standard for Quality Systems.

JIS–Japan Industrial Standards.

JUSE–Japanese Union of Scientists and Engineers.

Kaizen–Taken from the Japanese words kai and zen, where kai means **change** and zen means **good.** The popular meaning is continual improvement of all areas of a company not just quality.

LCL–Lower Control Limit (search in Control limit).

Lot or batch–One or more components or finished devices that consist of a single type, model, class, size, composition, or software version that are manufactured under essentially the same conditions and that are intended to have uniform characteristics and quality within specified limits.

LSL–Lower Specification Limit (search in Specification).

Maintainability–1. The probability that a failed system can be made operable in a specified interval or downtime.

2. Ability of an item under stated conditions of use to be retained in, or restored to, within a given period of time, a specified state in which it can perform its required functions when maintenance is performed under stated conditions and while using prescribed procedures and resources.

Management–Coordinated activities to direct and control an organization.

Management system–System to establish policy and objectives and to achieve those objectives.

MBNQA–Malcolm Baldrige National Quality Award.

Measurement control system–Set of interrelated or interacting elements necessary to achieve metrological confirmation and continual control of measurement processes.

Measurement process–Set of operations to determine the value of a quantity.

Measuring equipment–Measuring instrument, software, measurement standard, reference material or auxiliary apparatus or combination thereof necessary to realize a measurement process.

Metrological characteristic–Distinguishing feature which can influence the results of measurement.

Metrological confirmation–Set of operations required to ensure that measuring equipment conforms to the requirements for its intended use.

Metrological function–Function with organizational responsibility for defining and implementing the measurement control system.

MRB–Material review board

MSA–Measurement System Analysis

MTBF–Mean time between failures

NACCB–National Accreditation Council for Certification Bodies (UK)

NDT–Non-destructive testing

NIST–National Institute of Science and Technology

Nonconformance–Product or material which does not conform to the customer requirements or specifications.

Nonconformities–Specific occurrences of a condition that does not conform to specifications or other inspection standards; sometimes called discrepancies or defects

Nonconformity–A process which does not conform to a quality system requirement.

Normal Distribution–Search in Distribution

Numerical reliability–The probability that an item will perform a required function under stated conditions for a stated period of time. (Search in MTBF)

Objective evidence–Data supporting the existence or verity of something.

Organization–Group of people and facilities with an arrangement of responsibilities, authorities and relationships.

Organizational structure–Arrangement of responsibilities, authorities and relationships between people.

Parts Per Million (PPM)–PPM is a way of stating the performance of a process in terms of actual or projected defective material.

PFMEA–Process Failure Mode and Effects Analysis

Population–The universe of data under investigation from which a sample will be taken.

Prevention–A future-oriented strategy that improves quality by directing analysis and action toward correcting the production process. Prevention is consistent with a philosophy of never-ending improvement.

Preventive action–Action to eliminate the cause of a potential nonconformity or other undesirable potential situation.

Procedure–Specified way to carry out an activity or a process.

Process–1. The combination of people, machine and equipment, raw materials, methods, and environment that produces a given product or service.

2. Set of interrelated or interacting activities which transforms inputs into outputs.

Process Capability–The measured, built-in reproducibility (consistency) of the product turned out by the process. Such a determination is made using statistical methods, not wishful thinking. The statistically determined pattern or distribution can only then be compared to specification limits to decide if a process can consistently deliver product within those parameters.

Process Failure Mode and Effects Analysis (PFMEA)–An analytical technique used by a manufacturing responsible engineer/team as a means to assure that, to the extent possible, potential failure modes and their associated causes/mechanisms have been considered and addressed.

Process quality audit–An analysis of elements of a process and appraisal of completeness, correctness of conditions, and probable effectiveness.

Process validation–Establishing by objective evidence that a process consistently produces a result or product meeting its predetermined specifications.

Product–Result of a process. (May be services, software, hardware or processed materials, or a combination thereof.)

Product liability or Service liability–A generic term used to describe the onus on a producer or others to make restitution for loss related to personal injury, property damage, or other harm caused by a product or service.

Product quality audit–A quantitative assessment of conformance to required product characteristics.

Project–Unique process, consisting of a set of coordinated and controlled activities with start and finish dates, undertaken to achieve an objective conforming to specific requirements, including the constraints of time, cost and resources.

QFD–Quality Function Deployment.

QMS–Quality Management System (search in Quality system)

QS-9000–Quality system requirements for suppliers to Daimler Chrysler, Ford and General Motors

QSR–Quality System Requirements

Qualification process–Process to demonstrate the ability to fulfill specified requirements.

Quality–1. The totality of features and characteristics that bear on the ability of a device to satisfy fitness-for-use, including safety and performance.

2. Degree to which a set of inherent (existing) characteristics fulfils requirements.

Quality assurance–All those planned or systematic actions necessary to provide adequate confidence that a product or service will satisfy given requirements for quality.

Quality audit (also quality assessment, or conformity assessment)–A systematic and independent examination and evaluation to determine whether quality activities and results comply with planned arrangements and whether these arrangements are implemented effectively and are suitable to achieve objectives. (search in Quality system audit)

Quality characteristic–Inherent characteristic of a product, process or system related to a requirement.

Quality control–The operational techniques and the activities used to fulfill requirements of quality.

Quality engineering–That branch of engineering which deals with the principles and practice of product and service quality assurance and control.

Quality improvement–Part of quality management focused on increasing the ability to fulfill quality requirements.

Quality loop; Quality spiral–Conceptual model of interacting activities that influence the quality of a product or service in the various stages ranging from the identification of needs to the assessment of whether these needs have been satisfied.

Quality management–Coordinated activities to direct and control an organization with regard to quality.

Quality management system–Management system to direct and control an organization with regard to quality.

Quality manual–Document specifying the quality management system of an organization.

1. A quality manual may relate to the totality of an organization's activities or only to a part of it. The title and scope of the manual reflects the field of application.

2. A quality manual can vary in depth and format to suit the needs of an organization. It may be used, for example, "quality assurance manual." "quality management manual."

Quality measure–A quantitative measure of the features and characteristics of a product or service.

Quality objective–Something sought, or aimed for, related to quality.

Quality policy–1. The responsibilities, authorities, and interrelationships of personnel who manage, perform, verify, or review work affecting quality

2. The quality system procedures and instruction

3. Overall intentions and direction of an organization related to quality as formally expressed by top management.

Quality Plan–A document setting out the specific quality practices, resources, and activities relevant to a particular product, process, service, contract, or project.

Quality plan audit–*Search in* Quality system audit

Quality Planning–1. A structured process for defining the methods (*i.e.*, measurements, tests) that will be used in the production of a specific product or family of products (*i.e.*, parts, materials). (Search in Quality Plan)

2. Part of quality management focused on setting quality objectives and specifying necessary operational processes and related resources to fulfill the quality objectives.

Quality surveillance–The continuing monitoring and verification of the status of procedures, methods, conditions, products, processes, and services, and analysis of records in relation to stated references to ensure that requirements for quality are being met.

Quality system–The organizational structure, responsibilities, procedures, processes, and resources for implementing quality management.

Quality system audit–A documented activity performed to verify, by examination and evaluation of objective evidence, that applicable elements of the quality system are suitable and have been developed, documented, and effectively implemented in accordance with specified requirements.

Quality system review–A formal evaluation by management of the status and adequacy of the quality system in relation to quality policy and/or new objectives resulting from changing circumstances.

Range–A measure of the variation in a set of data. It is calculated by subtracting the lowest value in the data set from the highest value in that same set.

Reaction Plan–The action specified by a Control Plan when nonconforming product or process instability is identified.

Realization (as used in ISO 9000:2000)–The carrying out of an action or process to completion.

Record–Document stating results achieved or providing evidence of activities performed.

Recorder–The team member that takes minutes during team meetings to capture team's progress. Once the team is well underway, this role can be rotated through out the group.

Registrar–A company that conducts quality system assessments to the Quality System Requirements.

Regrade–Alteration of the grade of a nonconforming product in order to make it conform to requirements differing from the original ones.

Regression analysis–A statistical technique used to determine the best mathematical expression to describe the relationship between a response and independent variables.

Relative quality–Degree of excellence of a product or service.

Release–Permission to proceed to the next stage of a process.

Reliability–1. The probability that an item will continue to function at customer expectation levels at a measurement point, under specified environmental and duty cycle conditions.

2. The ability of an item to perform a required function under stated conditions.

Reliability engineering–That engineering function dealing with the principles and practices related to the design, specification, assessment, and achievement of product or system reliability requirements and involving aspects of prediction, evaluation, production, and demonstration.

Repair–1. Action taken on nonconforming product so that the product will fulfill the intended usage although the product may not conform to the original requirements.

2. Action on a nonconforming product to make it acceptable for the intended use.

Requirement–Need or expectation that is stated, generally implied or obligatory.

Review–Activity undertaken to determine the suitability, adequacy and effectiveness of the subject matter to achieve established objectives.

Rework–1. Action taken on nonconforming product so that it will meet the specified requirements.

2. Action on a nonconforming product to make it conform to the requirements.

Robust–The ability of a product or service to function appropriately regardless of external conditions and other uncontrollable factors.

Robust design–An approach to the planning of new products and services that harnesses Taguchi methods.

Run chart–Also known as a line chart, or line graph. A chart that plots data over time, allowing you to identify trends and anomalies.

Runs–The patterns in a Control Chart within which a number of points line up on only one side of the central line.

RvC–Raad voor de Certificatie (Dutch Council for Certification)

SAE–Society of Automotive Engineers

The engineering requirement for judging acceptability of a particular characteristic.

Sample–1. One or more individual events or measurements selected from the output of a process for purposes of identifying characteristics and performance of the whole.

2. A subset of a population used to represent the population in statistical analysis. Samples are almost always random, which means that all individuals in the population are equally likely to be chosen for the sample.

Sample standard deviation chart (s chart)–Control chart in which the standard deviation of the subgroup is tracked to determine the variation within a process over time. Sample standard deviation charts are usually paired with average charts for complete analysis.

Scatterplot–A tool that studies the possible relationship between two variables expressed on the x-axis and y-axis of a graph. The direction and density of the points plotted will indicate various relationships or a lack of any relationship between the variables.

SCC–Standards Council of Canada

Scrap–Action on a nonconforming product to preclude its originally intended use.

Seven tools of quality Quality improvement tools that include the histogram, Pareto chart, check sheet, control chart, cause-and-effect diagram, flowchart, and scatter diagram.

Shewhart cycle Another name for the Plan-Do-Check-Act cycle. It is also sometimes called the Deming cycle.

Sigma (σ)–The Greek letter used to designate the estimated standard deviation.

Simulation–The practice of mimicking some or all of the behavior of one system with a different, dissimilar system.

Simultaneous Engineering–A way of simultaneously designing products, and the processes for manufacturing those products, through the use of cross-functional teams to assure manufacturability and to reduce cycle time.

Six Sigma–Quality process, developed at Motorola, focused on reducing defects to a six sigma level (3.4 defects per million parts; 0.00034 per cent), for all practical purposes zero defects.

SMWT–Self-Managed Work Teams.

Software–An intellectual creation consisting of information expressed through supporting medium.

SPC–Statistical Process Control

Special Cause–A source of variation that is intermittent, unpredictable, unstable; sometimes called an assignable cause.

Special causes–Causes of variation in a process that are not inherent in the process itself but originate from circumstances that are out of the ordinary. Special causes are indicated by points that fall outside the limits of a control chart.

Specification limit An engineering or design requirement that must be met in order to produce a satisfactory product..

Specification–1. The document that prescribes the requirements with which the product or service has to conform.

2. The engineering requirement for judging acceptability of a particular characteristic. Chosen with respect to functional or customer requirements for the product, a specification may or may not be consistent with the demonstrated capability of the process (if it is not, out-of-specification parts are certain to be made). A specification should never be confused with a control limit.

SQC (Statistical Quality Control Stakeholder)–An individual or group of individuals with a common interest in the performance of the supplier organization and the environment in which it operates.

Stakeholder An individual or group of individuals with a common interest in the performance of the supplier organization and the environment in which it operates.

Standard Deviation–A measure of the spread of the process output or the spread of a sampling statistic from the process (*e.g.*, of subgroup averages), denoted by the Greek letter **σ** (sigma) for the estimated standard deviation. Search in Sigma

Statistical Control–The condition describing a process from which all special causes have been removed, evidenced on a control chart by the absence of points beyond the control limits and by the absence of non-random patterns or trends within the control limits.

Statistical process control (SPC)–1. Analysis and control of a process through the use of statistical techniques, particularly control charts.

2. The use of statistical techniques such as Control Charts to analyze a process or its output to take appropriate actions to achieve and maintain a state of statistical control and to improve the capability of the process.

Statistical quality control (SQC)–Analysis and control of quality through the use of statistical techniques, essentially the same as SPC.

Stratification–The process of classifying data into subgroups based on characteristics or categories.

Structural variation–Variation caused by recurring system-wide changes such as seasonal changes or long-term trends.

Supplier–1. Organization or person that provides a product.

2. Anyone whose output (materials, information, service, etc.) becomes an input to another person or group in a process of work. A supplier can be external or internal to the organization.

System–Set of interrelated or interacting elements.

Taguchi, Genichi–Developed a set of practices known as Taguchi Methods, as they are known in the U.S., for improving quality while reducing costs. Taguchi Methods focus on the design of efficient experiments, and the increasing of signal to noise ratios. Dr. Taguchi also articulated the developed the quality loss function. Currently, he is executive director of the American Supplier Institute and director of the Japan Industrial Technology Institute.

Tampering–Dr. Deming cautions against tampering with systems that are "in control." It is very common for management to react to variation which is in fact normal, thereby starting wild goose chases after sources of problems which don't exist. Tampering with stable processes actually increases variation.

Technical expert (in an audit)–Person who provides specific knowledge of or expertise on the subject to be audited.

Test–Determination of one or more characteristics according to a procedure.

Testing–A means of determining the capability of an item to meet specified requirements by subjecting the item to a set of physical, chemical, environmental, or operating actions and conditions.

TGA–Germany Association for Accreditation

Timekeeper–Team member who keeps track of time spent on each agenda item during team meetings. This job can easily be rotated among team members.

TL 9000–Quality system requirements for suppliers to the telecommunications industry

Top management–Person or group of people who directs and controls an organization at the highest level.

Total Quality Management–Managing for quality in all aspects of an organization focusing on employee participation and customer satisfaction. Often used as a catch-all phrase for implementing various quality control and improvement tools.

TPM–Total Productive Maintenance

TQM–Total Quality Management

Traceability–The ability to trace the history, application, or location of an item or activity and like items or activities by means of recorded identification.

Tree diagram–A chart used to break any task, goal, or category into increasingly detailed levels of information. Family trees are the classic example of a tree diagram. In PathMaker, the structure of the tree diagram is identical to that of the cause and effect diagram.

Type I error–Rejecting something that is acceptable. Also known as an alpha error.

Type II error–Accepting something that should have been rejected. Also known as beta error.

U chart–A control chart showing the count of defects per unit in a series of random samples.

UCL–Upper control limit (search in Control limit)

USL–Upper specification limit (search in Specification)

Validation–Confirmation by examination and provision of objective evidence that the particular requirements for a specific intended use can be consistently fulfilled.

1. In design and development, validation concerns the process of examining a product to determine conformity with user needs.

2. Validation normally performed on the final product under defined operating conditions it may be necessary in earlier stages.

3. The term "validated" is used to designate the corresponding status.

4. Multiple validation may be carried out if there are different intended uses.

Variables–Those characteristics of a part that can be measured. Examples are length in millimeters, resistance in ohms, closing effort of a door in kilograms, and the torque of a nut in foot pounds. (Search in Attributes)

Value added–Each time work is done to inputs to transform them into something of greater usefulness as an end product.

Variables data–Data that is measured on a continuous and infinite scale such as temperature, distance, and pressure rather than in discreet units or yes/no options. Variables data is used to create histograms, some control charts, and sometimes run charts.

Variables–Those characteristics of a part that can be measured. Examples are length in millimeters, resistance in ohms, closing effort of a door in kilograms, and the torque of a nut in foot pounds. (Search in Attributes)

Variance–A measure of deviation from the mean in a sample or population.

Variation–Change in the output or result of a process. Variation can be caused by common causes, special causes, tampering, or structural variation.

Variation–The inevitable difference among individual outputs of a process. The sources of variation can be grouped into two major classes: Common Causes and Special Causes.

Verification–1.The act of reviewing, inspecting, testing, checking, auditing, or otherwise establishing and documenting whether items, processes, services, or documents conform to specified requirements.

2. Confirmation by examination and provision of objective evidence that specified requirements have been fulfilled.

Vision–Often incorporated into an organizational mission (or vision) statement to clarify what the organization hopes to be doing at some point in the future. The vision should act as a guide in choosing courses of action for the organization.

Voice of the Customer–Customer feedback both positive and negative; including likes, dislikes, problems and suggestions.

Voice of the Process–Statistical data that is feedback to the people in the process to make decisions about the process stability and/or capability as a tool for continual improvement. Search in Statistical Process Control)

Waiver–Written authorization to use or release a quantity of material, components, or stores already manufactured but not conforming to the specified requirements.

Walter A. Shewhart–The father of statistical process control or statistical quality control. He pioneered statistical quality control and improvement methods when he worked for Western Electric and Bell Telephone in the early decades of the 20th century.

Work environment–Set of conditions under which work is performed.

Zero defects–Philip Crosby's recommended performance standard that leaves no doubt regarding the goal of total quality. Crosby's theory holds that people can continually move closer to this goal by committing themselves to their work and the improvement process.

Zero Defects–The quality concept of zero tolerance for defects (search in Six Sigma).

Chapter 12
Crop Science Terms

Aberrant–Different from the normal type of species, genus, or higher group in one or more characters, but not readily assignable to another group.

Absolute growth rate–*Search in* growth rate.

Acceptability, animal–Readiness with which animals select and ingest a forage; sometimes used interchangeably to mean either palatability or voluntary intake.

Achlamydeous-Of flowers: Lacking petals and sepals; *e.g.*, willow (*Salix* spp.).

Acid-detergent fiber–Insoluble residue following extraction with acid detergent (van Soest method); *i.e.*, cell wall constituents minus hemicellulose. *Abbr.*: ADF.see *also* neutral-detergent fiber.

Acid-detergent lignin–Lignin determined on the residue following extraction with acid detergent. *Abbr.*: ADL.

Ad libitum feeding–Daily feed offerings in excess of consumption; generally, 115 per cent of consumption.

ADP–Adenosine 5'-diphosphate.

Aeration, mechanical–*Search in* cultivation, turf.

Aflatoxin ($C_{17}H_{10}O_6$)–A polynuclear substance and a known carcinogen derived from molds; specifically, that produced by a fungus occurring on peanut (Groundnut).

Agamospermy–All types of apomixes in which seeds are formed by asexual means. Does not include vegetative reproduction.

Agricultural holding–An agricultural holding is the economic unit under a single management engaged in agricultural production activities. The unit may also be engaged in non-agricultural activities so that this concept should not be interpreted too strictly; the aim is rather to value the final production of all agricultural products. Also, establishments or specialised units which provide agricultural services on a fee or contract basis should, in general, be included.

Agricultural land–Agricultural land is land including arable land, land under permanent crops and land under permanent meadows and pastures.

Agricultural pollution–Agricultural pollution comprises liquid and solid wastes from all types of farming activities, including run-off from Agrochemicals pesticide and fertilizer use, and from feedlots; erosion and dust from ploughing; animal manure and carcasses; and crop residues and debris.

Agricultural waste–Agricultural waste is waste produced as a result of various agricultural operations. It includes manure and other wastes from farms, poultry houses and slaughterhouses; harvest waste; fertilizer run-off from fields; pesticides that enter into water, air or soils; and salt and silt drained from fields.

Agricultural water use–Agricultural water use includes water abstracted from surface and groundwater, and return flows (withdrawals) from irrigation for some countries, but excludes precipitation directly onto agricultural land.

Agro-chemical–Agro-chemicals are commercially produced, usually synthetic, chemical compounds used in farming–such as a fertiliser, pesticide or soil conditioner.

Agro-ecology–Agro-ecology is the study of the relation of agricultural crops and environment

Agro-ecosystem–An agro-ecosystem is an ecosystem under agricultural management, connected to other ecosystems.

Agroforestry–1. Land-use system in which woody perennials are grown for wood production with agricultural crops, with or without animal production.

2. Agro-forestry is the collective term used for land–use systems and technologies in which woody perennials (trees, shrubs, palms, bamboo's and so forth) are deliberately used on the same land management unit as agricultural crops and / or animals, in some form of either spatial arrangement or temporal sequence

Agronomy: A specialization of agriculture concerned with the theory and practice of field crop management.

Agrosilviculture–An agroforestry system in which land is used for agricultural and forest crops.

Agrosilvipasture–An agroforestry system in which land is used for agriculture, forestry and animal husbandry in various spatial and temporal arrangements.

Aleurone–A layer of protein granules found in the endosperm of ripe seeds or fruits.

Alkaline hydrolysis–Cleavage of covalent chemical bonds by high pH. RNA molecules have 2' hydroxyls that attack the 3' phosphodiester bond at high pH and degrade the RNA polymer to monoribonucleotides. DNA lacks 2' hydroxyl groups and is not hydrolyzed at high pH.

Allele–Any of several alternative forms of a gene. Allels of a gene always occupy some site (locus) on homologous chromosomes. see *also* locus.

Allometric–A relationship between two variables of the form $y = ax^b$ where a and b are constants such that $\ln y = b\ln x + \ln a$; b is termed the allometric coefficient. Of common use in morphogenesis studies, where organ dimensions and growth rates frequently can be fit by the relation.

Alloplasm–Cytoplasm from an alien species that has been transferred by backcrossing into a cultivated species. The alien cytoplasm's most common and readily observable effects are male sterility, female sterility, or reduced plant vigor. If the alien cytoplasm does not manifest itself, the cytoplasm is not considered alien.

Allozyme–Alternative form of an enzyme encoded by different alleles at a given locus. *Compare* isoenzyme.

Amendment, physical–Any substance (such as sand, calcined clay, peat, or sawdust) added to soil for the purpose of altering physical conditions.

Amphidiploid–A plant possessing the sum of the somatic chromosome number of two species.

Amphiphloic–Having both internal and external phloem.

Amphivasal bundle–Concentric vascular bundle in which the xylem surrounds the phloem, as in the rhizomes of some monocots. *Compare* amphicribral bundle.

Amplification–Process by which the copy number of a gene or plasmid is increased.

Androdioecious–Plants that have male and perfect flowers on separate plants. *Compare* andromonoecious.

Andromonoecious–A plant bearing bisexual or perfect flowers in addition to staminate flowers. *Compare* androdioecious.

Aneuploid–(1) Characterizing a cell in which the nucleus does not contain an exact multiple of the haploid number of chromosomes, one or more chromosomes being present in a greater or lesser number than the rest. The chromosomes may or may not show rearrangements. (2) Such a cell or organism. *See*–ploid.

Animal day–One day's tenure upon a pasture by one animal. *Usage:* Not synonymous with animal-unit day.

Animal month–One month's tenure upon pasture by one animal. *Usage:* Not synonymous with animal-unit month.

Animal unit–One mature nonlactating bovine weighing 500 kg and fed at maintenance level, or, in other kinds of animals, the equivalent, expressed as $(weight)^{0.75}$.

Anisogamy–Sexual reproduction in which the female gamete is large and immobile and the male gametes is small and motile. *Compare* oogamy.see *also* autogamy.

Anther culture–The in vitro culturing, on a synthetic medium, of anthers containing microspores. The microspores may form haploid callus or develop directly into haploid plants.see androgenesis.

Anthesis–The action or period of opening of a flower, especially the period when the stigma is favorable for the germination of pollen coming in contact with it.see *also* blooming; flowering.

Anthoxanthins–Water-soluble pigments, widely distributed in leaves, stems, and flowers. They are mostly colorless, but can modify other colors, especially anthocyanins.

Antibiosis–Physiological antagonism of one organism toward another, used chiefly in reference to such antagonism among fungi.see *also* resistance and related entries.

Anticodon–A specific sequence of three ribonucleotides in a tRNA molecule that base-pair with the three complementary nucleotides of the codon sequence within mRNA molecules. This codon-anticodon interaction occurs on ribosomes and specifies the insertion of amino acids in the correct order during elongation of polypeptides.

Antioxidant–An organic compound that accepts free radicals and thus prevents autoxidation of fats and oils. At very low concentrations in food, antioxidants not only act in retarding rancidity but protect the nutritional value or minimize the breakdown of vitamins and essential fatty acids.

Antisense gene–An engineered gene placed in inverted orientation relative to a promoter that when transcribed, produces a transcript complementary to the mRNA transcribed from the normal orientation of the gene.

Antiserum–Blood serum containing specific antibodies against an antigen.

Apoplast–The continuum of nonliving plant cell-wall material that surrounds the symplast. It includes the tracheary elements of xylem and the free space, and is one of two parallel pathways for solute movement through plants (the other being symplast). Substances can move through the apoplast only down a thermodynamic potential gradient.

Apparent free space–A measure of free space gained from ion influx and efflux experiments; it is influenced by ion exchange of solutes with cell walls and thus can differ for each solute. *Abbr.*: AFS.

Aquaculture–Aquaculture is the farming of aquatic organisms including fish, mollusks, crustaceans and aquatic plants. Farming implies some sort of intervention in the rearing process to enhance production, such as regular stocking, feeding, protection from predators and so forth. It also implies individual or corporate ownership of the stock being cultivated.

Aroma–A characteristic odor, as of a plant, feed, or food.

Artificial turf–A fabricated rug of fibers simulating turf. *Usage:* Astroturf is a trademark, not the general term.

Aseptic–Characterized by absence of contaminating agents like fungi, bacteria, viruses, mycoplasms, and other microorganisms (*e.g.*, in cultures).

Ash–(1) The residue remaining after complete burning of combustible matter. (2) *See* cell-wall constituents.

Assimilation–(1) The incorporation of inorganic forms of chemical elements into organic forms termed *assimilates*. Thus, we have the assimilation of CO_2 into photosynthates such as sugar and of NO_3, NH_4, and SO_4 into amino acids. (2) In growth analysis, net assimilation rate is given a specific definition as the rate of change of crop biomass (*i.e.*, crop growth rate, reflecting the net of photosynthetic and respiratory activities) per unit leaf area index. Net assimilation rates can also be defined for carbon, nitrogen, or other entities. (3) The formation of protoplasm.

Atavism–Recurrence of an ancestral character after an interval of several generations.

Atmosphere–The atmosphere is the mass of air surrounding the earth, composed largely of oxygen and nitrogen.

Atmospheric absorption–Atmospheric absorption is absorption by the earth's atmosphere of most of the X-rays and ultraviolet and infrared radiation emitted by the sun, except visible light. It prevents the earth's surface from becoming too hot.

Autecology–The ecology of individual species, as opposed to that of community ecology.

Autonomic movement–Plant movement as a result of internal stimuli. *Synonym:* autogenic or spontaneous movement.

Autonomous–*Search in* habituation.

Autoradiography–A method for determining the presence and location of radioactively labeled molecules by their effect in creating an image on a photographic emulsion, usually x-ray film, by activating the silver halide grains, which are then reduced to metallic silver when the film is developed. Autoradiography is commonly used to determine whether a radioactive probe molecule has hybridized to denatured DNA or RNA following Southern or northern transfers, respectively, or in colony hybridization procedures.

Autotrophic–Self-reliant organisms synthesizing their own food like green plants, through converting environmental energy into chemical energy. *Compare* heterotrophic.

Auxotroph–A mutant requiring specific compound(s) for its growth that are already present in the wildtype. Useful as a genetic marker and for complementation tests. *adj.* auxotrophic. *Compare* autotroph.

Available pasture–*Usage:* Not a recommended term (pasture refers to a specific type of grazing management unit, not to that which is consumed).see forage mass; available forage.

Avoidance–Ability of an organism to prevent an injurious stress, pathogen, or predator from penetrating its tissues (*e.g.*, drought avoidance may be achieved through restriction of water loss or by expansion of the root system to a greater supply of water).

Axenic–Without other living organisms. Used to refer to a pure culture, *i.e.*, a culture uncontaminated by symbionts or parasites.

Backgrounding–Intensive management of young cattle (post weaning) to facilitate maximum performance.

Bacteriophage–A virus that infects a bacterium. Bacteriophages are widely applied in recombinant DNA technology. *Usage:* Usually shortened to phage.

Balanced diallelic–The genotype involving a multiple allelic locus in an autotetraploid where two different alleles are represented an equal number of times (*e.g.*, $a_1a_1a_2a_2$).

Ball roll–The distance a ball moves (i) after striking the ground upon termination of its air flight, (ii) as a result of a putting stroke, or (iii) as a result of hand-imparted motion, as in lawn bowling.

BAP–(1) Bacterial alkaline phosphatase, the enzyme from *Escherichia coli* that is used to remove 5' terminal phosphate groups from DNA chains to prevent recircularization of vector molecules during gene cloning experiments. (2) Benzylaminopurine, a cytokinin.

Bare fallow–Complete inversion and incorporation of residues for maximum decomposition, done to prevent the growth of all vegetation; usually associated with summer fallow.

Base pair–Two nucleotide bases on different strands of the nucleic acid molecule that form hydrogen bonds between them. The bases can pair in a single way only: adenine with thymine (in DNA) or uracil (in RNA), and guanine with cytosine.

Batch drying–Drying seed in relatively small quantities held in a stationary position, as opposed to drying in a continuous moving line.

Bench setting–Height that the cutting plane of a mower is set above a hard, level surface.

Bicollateral vascular bundle–A vascular bundle with phloem on both sides of the xylem instead of on the outer side only.

Bioassay–(1) The use of living organisms to quantitatively estimate the amount of biologically active substances present in a sample. (2) In cell biology and molecular genetics: determination of the effectiveness of a compound by measuring its effect on plants or animal tissues or organisms in comparison with a standard preparation. Extensively used to measure effects of different auxins and gibberellins and auxins or pollutants on plants.

Biodiversity–Biodiversity refers to the range of genetic differences, species differences and ecosystem differences in a given area.

Biodiversity indices–Biodiversity indices are measures of species diversity expressed as ratios between numbers of species and "importance values" (numbers, biomass, productivity and so on) of individuals. The term may also refer to genetic diversity and diversity of habitats or communities.

Biodegradation–Biodegradation is the process by which organic substances are decomposed by micro-organisms (mainly aerobic bacteria) into simpler substances such as carbon dioxide, water and ammonia.

Bio-gas–A combustible gas (composed primarily of methane) produced when sewage or manure is fermented in the absence of oxygen. The solid material that remains in the digester after fermentation can be used as an organic fertilizer.

Biological accumulation–Biological accumulation is the accumulation of elements and compounds of harmful substances in the tissues of living organisms.

Biological containment–A strategy to reduce the risks of recombinant molecules propagated within microorganisms being released into the general environment.

Biological diversity–Biological diversity refers to the global variety of species and ecosystems and the ecological processes of which they are part, covering three components: genetic, species and ecosystem diversity.

Biological pest control–Biological pest control refers to the use of predatory or parasitic organisms instead of highly polluting chemicals to reduce the number of harmful animals or plants, as in, for example, the destruction of the citrophilus mealy bug by parasitic species of the chalcid wasp; the predation of beetles on the cottony—cushion scale; and the control of Japanese beetles by Bacillus popilliae.

Biomass–1. The weight of living organisms (plants and animals) in an ecosystem, at a given point in time, expressed as fresh or dry weight.

2. Biomass is the quantity of living material of plant or animal origin, present at a given time within a given area.

Bioreactor–Culture vessel used for large-scale bioprocessing.

Biotechnology–A wide range of techniques used to manipulate living organisms to develop or accentuate characteristics that human beings desire. Genetic engineering, in which the hereditary material of a plant or an animal modified at the molecular level, is one contemporary form of biotechnology.

Biotinylated DNA–DNA labeled with biotin by incorporation of biotinylated dUTP. It is used in place of radioactive probes in hybridization analyses. Hybridization is detected by use of a complex of streptavidin-biotin-horseradish peroxidase that produces green fluorescence at the location of hybridization.

Bird resistance–A characteristic of a genotype or individual plant in which it is avoided by birds until other food sources are exhausted, or until the plant is weathered.

Blend, seed–A combination ofseed of varieties and/or composite-cross populations of the same kind, each present in excess of 5 per cent by weight of the whole, and derived by growing the component varieties and/or composite-cross populations separately and mixingseed to form the commercial class of seed.

Bloom–Of sorghum: Characterized by the presence of a white, waxy, or pruinose (*i.e.*, a whitish dust) covering on the leaves and stems by a high density of planktonic organisms. *Antonym:* bloomless.

Bloom, full–The developmental stage at which essentially all florets in the inflorescence are in anthesis.

Blooming–In the grass family: the period during which florets are open and anthers are extended. *Synonym:* anthesis.

Blot–(1) To transfer DNA, RNA, or protein, usually from an electrophoretic gel to an immobilizing matrix such as diazobenzyloxymethyl (DBM) paper, nitrocellulose, or nylon membranes. (2) The autoradiograph obtained from hybridization analysis after the transfer.see *also* electroblotting; northern blot; Southern blot; western blot.

Body weight, fat-free–Conceptually, in livestock, empty body weight minus body fat; in practice, generally determined from the ratio of carcass specific gravity to body specific gravity.

Bollsize–Weight in grams ofseed cotton from one boll.

Bomb calorimetry–Process whereby a substance is completely oxidized in 25 to 30 atmospheres (~2.5 to 3.0 MPa) of oxygen (O_2) to determine gross energy content.

Botanical variety–An infraspecific taxon in botanical nomenclature, below the rank of subspecies. *Abbr.*: var. *Usage:* The abbreviation in roman type; the name, in italics; no capitals. *Compare* cultivar.

Bp–The symbol for base pair, a measure of the size of a double-stranded nucleic acid. *Usage:* The SI prefix k may be used, but the symbol changes: 1000 bp = 1 kb (not 1 kbp).

Bran–Pericarp of grain.

Breaking load–The maximum force applied to a structure in a tensile test carried to rupture.

Broadcast sprigging–Vegetative turf establishment by broadcasting and covering of stolons, rhizomes, or tillers with soil.

Browning–The reaction between reducing sugars and free amino groups in proteins to form a complex that undergoes a series of reactions to produce brown polymers usually referred to as melanoidins. Higher temperatures and basic pH favor the reaction. The process is desirable for flavor and aroma in certain food processing situations (*e.g.*, toasting bread) but is detrimental in other situations (*e.g.*, dryingseed and food products).

Browse–(1) Leaf and twig growth of shrubs, woody vines, trees, cacti, and other nonherbaceous vegetation available for animal consumption. (2) *v.* To consume browse in situ (of animals). *Compare* forage; graze.

Bunchgrass–*Search in* bunch-type growth habit.

Calcined clay–Clay minerals, such as montmorillonite and attapulgite, that have been fired at high temperatures to obtain absorbent, stable, granular particles; used as amendments in soil modification.

Callus–1. Undifferentiated tissue cultured in vitro on a synthetic medium.

2. Mass of thin-walled cells usually developed as the results of wounding or in tissue culture.

Calorie–The amount of heat energy required to raise the temperature of 1 g of H_2O by 1 °C. *Synonym:* gram calorie. *Symbol:* cal. 1 kilocalorie (kcal) = 1000 calories; 1 megacalorie (Mcal) = 1 000 000 calories. The SI equivalent is the joule (1 calorie ~ 4.19 J).

Canopy–The vertical projection downward of the aerial portion of plants, usually expressed as percent of ground so occupied.

Cap site–The probable site of transcription initiation for eukaryotic genes. The primary transcripts of most eukaryotic mRNAs have an A (adenine) in the first position and the cap is added 5' to it.

Carbohydrates, structural–Carbohydrates found in the cell walls (*e.g.*, hemicellulose, cellulose).

Carding–The process of untangling and partially straightening fibers by passing them between two closely spaced surfaces, at least one of which is covered with sharp points that move at different speeds, thus converting a tangled mass of fibers to a filmy web.

Carrying capacity–The maximum stocking rate that will achieve a target level of animal performance, in a specified grazing method, that can be applied over a defined period of time without deterioration of the ecosystem. *Usage:* Carrying capacity is not static from season to season or year to year and may be defined over fractional parts of years. Average carrying capacity refers to long-term carrying capacity averaged over years; annual carrying capacity refers to a specific year. *Synonym:* grazing capacity.

Cash cropping–Growing crops for monetary return rather than direct food supply.

Castings, earthworm–Soil and plant remains excreted and deposited by earthworms in or on the turf surface or in their burrows. Castings form relatively stable soil granules that can, on closely mowed turf, produce an uneven surface. *Synonym:* earthworm casts.

Catchment–A drainage basin, or the area drained by a particular river system. Adjacent drainage basins are separated by watersheds. In North America, the term watershed refers to the entire drainage basin, and the height of land between basins referred to as a divide.

C-banding–*Centromere banding*–(1) Technique for chromosome identification, using modified Giemsa staining to reveal the distribution of constitutive heterochromatin, expressed at principal stages of meiotic and mitotic division. (2) The banding pattern thus revealed.see *also* G-banding; N-banding; Q-banding; R-banding.

CCC–*Search in* covalently closed circle.

cDNA cloning–The use of cDNA to clone the coding sequence of a gene or sets of genes by starting with mRNA. The cDNA from mature processed transcripts does not contain introns, the presence of which could act to prevent expression of plant or animal cDNAs when attached to a suitable promoter and cloned into a host organism such as bacteria or yeast.

Cell culture–The growing of cells in vitro, including the culture of single cells or suspensions of small aggregates of cells. In cell cultures, the cells are no longer organized into tissues. *Compare* tissue culture. see *also* callus; suspension culture.

Cell generation time–The interval between consecutive divisions of a cell. This interval can best be determined, at present, with the aid of cinephotomicrography.

Cell line–(1) Descent, through cell division, from a single original cell. (2) More specifically: A cell line arises from a primary culture at the time of the first successful subculture. The term implies that cultures from a cell line consist of numerous lineages of cells originally present in the primary culture.

Cell-free transcription–*Search in* in vitro transcription.

Cellulose–A carbohydrate formed from glucose, and a major constituent of plant cell walls; a colorless solid, insoluble in water. *Compare* hemicellulose.

Cell-wall constituents–Compounds that make up or constitute the cell wall, including cellulose, hemicellulose, lignin, and minerals (ash); *i.e.*, neutral-detergent fiber.

Center of origin–The geographic region in which a taxon (species, genus, family, etc.) or a crop originates.

Centrifugal spreader–An applicator from which dry particulate material is broadcast as it drops onto a spinning disk or blade beneath the hopper.

Certified seed–Certified seed is the progeny of breeder, select, foundation, or registeredseed so handled as to maintain satisfactory genetic purity and identity, and which has been acceptable to the certifying agency. Certified treeseed is defined asseed from trees of proven genetic superiority, as defined by the certifying agency, produced so as to assure genetic identity seed from interspecific hybrids or forest trees may be included.

Cesium sulfate (Cs_2SO_4)–A cesium salt that forms solutions of greater density than cesium chloride (CsCl) and is used for isopycnic or density-gradient centrifugation of double-stranded RNA molecules.

Chain terminator–The UAA, UAG, and UGA codons (stop or termination codons) that do not code for an amino acid but rather signal ribosomes to terminate protein synthesis. Two stop codons may often be found together at the end of a coding sequence of RNA.

Chalky–Color descriptor (related to hardness) characterizing kernel endosperm of cereal grains, especially rice. A rice kernel with a high level of chalk is generally undesirable. The chalky appearance arises from the structure of the endosperm, which has many small voids between the starch granules and the protein matrix. These voids cause light to be refracted and hence the endosperm appears white to reflected light and opaque to transmitted light. *Synonyms:* floury; soft endosperm. *Compare:* corneous; vitreous; flinty.

Chemical tillage–*Search in* chemical fallow.

Chilling injury–Damage to plants at low temperatures in the absence of freezing. Chilling injury is common in plants of tropical or subtropical origin at temperatures <10 °C (<50 °F) or, for some fruits, <15 °C (<59 °F). The cause usually is a change in viscosity of lipids in membranes.

Chimera–(1) A recombinant DNA molecule that contains sequences from more than one organism. (2) At the plant level, a mixture of tissues of genetically different constitutions that have resulted from mutation, grafting, or gene transfer.

Chloroplast–A plastid containing chlorophyll and other photosynthetic pigments within which photosynthesis takes place.

Chromic oxide (Cr_2O_3)–A completely indigestible chemical used as an indicator to estimate forage intake.

Chromosome walking–The sequential isolation of clones carrying overlapping sequences of DNA. The technique allows large regions of a chromosome to be covered, and can be used to identify which clone in a gene bank contains a desired gene or sequence that cannot be easily selected directly. Initially, a known gene is used to identify a clone containing it in colony hybridizations and then to obtain adjacent sequences for a second round of hybridization to identify clones with even more distal sequences. By repeating this process, the distance between a known cloned gene and a second uncloned gene of interest can be spanned, provided that both genes are on the same chromosome and the gene bank contains clones with all the intervening sequences.

Cleanseed–*Usage:* For tall fescue seed, endophyte-free seed is the preferred term.

Cleared lysate–The intracellular contents obtained by cell disruption, detergent treatment, and centrifugation to remove cell debris and large particles. The resulting solution contains primarily nucleic acids and protein. Preparation of a cleared lysate is usually the first step in plasmid DNA purification.

Cline–(1) A character gradient within a species, related to geographic distribution. (2) A more or less uniform series of variants from one extreme to the other.

Clippings–Leaves, and in some cases stems, cut off by mowing.

Clonal propagation–Asexual reproduction of plants that are considered to be physiologically and/or genetically uniform and to have originated from a single individual or explant.

Clone–All the asexually produced progeny of a single original individual that have been propagated vegetatively, usually by cuttings or natural propagation of bulbs, tubers, or rhizomes.

Closed continuous culture–A continuous culture in which inflow of fresh medium is balanced by outflow of corresponding volumes of spent medium. Cells are separated mechanically from outflowing medium and added back to the culture.

Codex Alimentarius–The Codex Alimentarius is an international code for food developed and administered by the United Nations' Codex Alimentarius Commission. Sometimes simply referred to as "the Codex".

Codex Alimentarius Commission–The Codex Alimentarius Commission is an international body charged with developing the standards, guidelines and recommendations that comprise the Codex Alimentarius.

Coding sequence–That portion of a gene that directly specifies the amino acid sequence of its protein product.

Noncoding sequences of genes include control regions, such as promoters, operators, enhancers, terminators, and introns.

Codon–A sequence of three nucleotide bases that specifies an amino acid or represents a signal to start or stop translation. *Usage:* Write codon names without intervening punctuation or spaces (*e.g.*, UAU, UAG). *Compare* anticodon.see *also* start codon; termination codon.

Cold test–Aseed vigor test in whichseeds are planted for a period of time in cool, moist unsterilized soil before transfer to a higher temperature, designed to approximate unfavorable planting conditions.

Cold-water insoluble nitrogen–A form of fertilizer nitrogen not soluble in cold water (25 °C; 77 °F). *Abbr.*: WIN.see *also* nitrogen activity index.

Cold-water soluble nitrogen–A form of fertilizer nitrogen soluble in cold water (25 °C; 77 °F). *Abbr.*: WSN.see *also* nitrogen activity index.

Colorant–A dye, pigment, or paintlike material applied to turf to create a favorable green color when the grass is discolored or damaged.

Combing–Using a comb to lift stolons and procumbent shoots so that they can be cut by a mower; the comb typically has metal teeth or flexible tines and is fastened immediately in front of a reel mower.

Common Agricultural Policy (CAP)–The CAP came into force in 1962, and is a system of rules which regulate the production, trade, and processing of agricultural products in the European Union. It establishes a common market for agricultural products between member states, and protects those products from international competition.

Companion crop–One crop sown with another, used particularly of the small grains with which forage crops are sown. *Synonym* nurse crop. *Usage:* Companion crop is the preferred term.

Composite-cross population–A population generated by hybridizing more than two varieties and/or lines of normally self-fertilizing plants and propagating successive generations of the segregating population in bulk in specific environments so that natural selection is the principal force acting to produce genetic change. Artificial selection may also be imposed. The resulting population is expected to have a continuously changing genetic makeup. Breeder seed is not maintained as originally released.

Conditioned storage–Storage of seed under controlled conditions of temperature and relative humidity.

Conjugation–The transfer of genetic material between mating types of bacteria involving donors (F^+ type) and recipients (F^-).

Conservation (nature)–Protection against irreversible destruction and other undesirable changes, including the management of human use of organisms or ecosystems to ensure such use is sustainable.

Conservation tillage–Conservation tillage is a tillage system that creates a suitable soil environment for growing a crop and that conserves soil, water and energy resources mainly through the reduction in the intensity of tillage, and retention of plant residues.

Consumer acceptance–The likelihood that the consumer will use the genotype, phenotype, or product by choice.

Continuous culture–A suspension culture continuously supplied with nutrients by the inflow of fresh medium. The culture volume is normally constant. *see also* closed continuous culture; open continuous culture.

Continuous stocking–A method of grazing livestock on a given unit of land where animals have unrestricted and uninterrupted access throughout the time period when grazing is allowed. *Synonym* (not recomme-nded): continuous grazing.see *also* rotational stocking; set stocking.

Contour ploughing–A method of cultivation designed to reduce soil erosion. Ploughing is carried out across the slope, rather than up and down it. This reduces the flow of water and reduces the potential for erosion.

Conventional planting–Planting that takes place days to months after a fallow period; usually involves primary and numerous secondary tillage operations. *Synonym:* delayed planting.

Convergent improvement–A breeding method involving the reciprocal addition to each of two inbred lines of the dominant favorable genes lacking in one line and present in the other. Backcrossing and selection are performed in parallel, each of the original lines serving as the recurrent parent in one series.

Cool-season turfgrass–Turfgrass species best adapted to growth during cool, moist periods of the year, commonly having temperature optimums of 15 to 25 °C (60-75 °F).

Copy number–(1) The number of plasmid molecules per cell. (2) The number of times a specific gene is present per haploid genome.see *also* low-copy; multicopy; single-copy.

Cored hay samples–Samples taken from stored hay, using a hollow cylinder to remove a core.

Corneous–Characterizing hard, vitreous, or horny endosperm in cereal grains.see *also* chalky.

Cost-benefit analysis–Cost-benefit analysis refers to the assessment of direct economic and social costs and benefits of a proposed project for the purpose of project or programme selection.

Coumestrol–Estrogenic compound occurring naturally in forage crops, especially in ladino clover, strawberry clover, and alfalfa.

Covalently closed circle–A double-stranded DNA molecule with no free ends that has a naturally supercoiled configuration. The two strands are interlinked and remain together even after denaturation if there are no single-strand nicks. *Abbr.:* CCC.

Cover crop–A cover crop is a temporary vegetative cover that is grown to provide protection for the soil and the establishment of plants, particularly those which are slow growing.

Creeping growth habit–Plant development by extravaginal stem growth at or near the soil surface, with lateral spreading by rhizomes and/ or stolons. *Synonym:* spreading growth habit.

Critical level–As applied to plant response functions, that level of a factor which is just limiting to plant performance; sometimes specifically defined as 95 per cent of the performance when the factor is nonlimiting. *Usage:* Widely used in plant and soil analysis for evidence of nutrient deficiency or toxicity. With leaf-area index, used as the index sufficient for near-maximum light interception or growth rate.

Crop residue–Portion of plants remaining after seed harvest; refers mainly to grain crop residue, such as corn stover, or of small-grain straw and stubble.

Crop rotation–Arable system in which a field is planted with different crops in a regular sequence over a set period of years. For example, a corn crop, which has high nitrogen demands, might be followed by a leguminous crop like peas or clover to restore the nitrogen to the soil. Crop rotation helps replenish nutrients and control pest infestations.

Crop year–A crop year is a twelve-month period used for collecting data on a particular crop — generally corresponding to the natural planting and marketing cycle for that crop. Usually, a crop year begins in a month other than January.

Crop year, coarse grains–Crop year, coarse grains refers to the crop marketing year beginning 1 April for Japan, 1 July for the European Union and New Zealand, 1 August for Canada and 1 October for Australia. The US crop year begins 1 June for barley and oats and 1 September for maize and sorghum.

Crop year, oilseeds–Crop year, oilseeds refers to the crop marketing year beginning 1 April for Japan, 1 July for the European Union and New Zealand, 1 August for Canada and 1 October for Australia. The US crop year begins 1 June for rapeseed, 1 September for soyabeans and for sunflower seed.

Crop year, rice–Crop year, rice refers to the crop marketing year beginning 1 April for Japan, Australia, 1 August for the United States, 1 September for the European Union, 1 October for Mexico, 1 November for Korea and 1 January for other countries.

Crop year, wheat–Crop year, wheat refers to the crop marketing year beginning 1 April for Japan, 1 June for the United States, 1 July for the European Union and New Zealand, 1 August for Canada and 1 October for Australia.

Crossing over–1. Exchange of genes between two paired chromosomes.

2. An exchange of corresponding segments of chromatids of homologous chromosomes, occurs during meiosis.

Cross-pollination–Pollen of one flower on one plant applied (artificially or naturally) to the stigma of a flower on another plant.

Crown gall–A tumor formed predominantly on the stems of broad-leaved plants when infected with *Agrobacterium tumefaciens* containing a Ti-plasmid. Part of the Ti-plasmid (T-DNA) is transferred into the genome of the affected plant cells due to action of the Ti-plasmid *vir* genes. Expression of T-DNA genes for auxin and cytokinin synthesis cause gall formation even in the absence of the bacterium. Whole plants can sometimes be regenerated from crown gall tissue and some of these still contain T-DNA.

Cubing–Process of forming hay into high-density cubes to facilitate transportation, storage, and feeding.

Cultivar–(1) A variety, strain, or race that has originated and persisted under cultivation or was specifically developed for the purpose of cultivation. (2) For cultivated plants, the equivalent of botanical variety, in accordance with the International Code of Nomenclature of Cultivated Plants-1980. *Usage:* Cultivar names are not italicized, and are indicated by single quotes at first use, or the word cultivar (but not both). The abbreviation cv. is properly used only with a binomial name: *Genus species* cv. cultivarname. Omit the abbreviation if single quotes are used: *Genus species* 'cultivarname'.

Culture medium–Any nutrient system for the artificial cultivation of bacteria or other cells; usually a complex mixture of organic and inorganic materials. *Plural:* media.

Cup cutter–A hollow cylinder with a sharpened lower edge used to cut the hole or cup in a green, or to replace small spots of damaged sod.

Cut–(1) To make a double-stranded break in DNA, usually with a Type II restriction endonuclease. *Usage:* 'Cut' is a popular term; the more specific term is cleave. (2) Such a break.

Cuticle–A waxy layer secreted by epidermal cells on the outer surface of plants.

Cutout In cotton: the occurrence of physiologically indeterminate growth.

Cyanogenesis–The release of hydrocyanic acid (HCN), in the process of chemical change. *adj.* Cyanogenetic, cyanogenic. *Usage:* Cyanogenetic is the preferred adjectival form.

Cytokinins–A class of plant growth regulators (chemically and functionally related to the natural hormone zeatin) that cause cell division, cell differentiation, shoot differentiation, etc. Cytokinins commonly used in tissue culture include kinetin, benzylaminopurine (BAP), and 2-isopentenyladenine (2-iP).

Cytoplasmic Inheritance–Traits controlled by organelle genes in the chloroplast or mitochondrial genomes; such traits usually express the phenotype of the maternal parent, although examples of biparental or paternal transmission of organelle genomes do occur.

DDT (Dichlorodiphenyl–trichloroethane)–A chlorinated hydrocarbon once widely used as a broad-spectrum insecticide. Introduced during the Second World-War as a delousing agent, it proved very effective against diseases such as malaria, yellow fever and typhus, which were spread by insects. Over the longer term, serious side affects became apparent. Being a broad-spectrum product, it killed beneficial insects as well pests and could accumulate in the environment for perhaps 20 years. Although not soluble in water, it was soluble in fat, which allowed it to migrate up the food chain, where it accumulated in the body tissue of the predators. In birds it caused the thinning of egg shells, seriously reducing the breeding success of some species. By the mid-1960s, DDT was found to be widespread in the fatty tissue of the human population, passed on from mother to child through breast milk. Although the link between DDT concentration and human health was not clear, its potential to cause serious ecological disruption was recognised, and it was eventually banned or had its use severely restricted in the developed world.

de novo–Literally, 'from a new [beginning]' (Latin). (1) With reference to plant regeneration or developmental processes: arising from unorganized cells or tissues or from prodetermined cells.see *also* adventitious. (2) In molecular biology: not requiring a primer.

Deferment–Postponement or delay of grazing on an area, to achieve a specific management objective (*e.g.*, to provide time for plant reproduction, establishment of new plants, restoration of plant vigor, a return to environmental conditions appropriate for grazing, or accumulation of forage for later use).

Deferred grazing–The deferment of grazing in a nonsystematic rotation with other land units.

Degermed corn–Grain (corn) from which the embryo (germ) has been removed.

Dehumidification–The process by which water is removed from a substance; used specifically of moisture vapor removed from air.

Denaturation–(1) The breakdown of the secondary or tertiary structure of a protein or nucleic acid by physical or chemical means. (2) More specifically, the conversion of a double-stranded nucleic acid to the single-stranded state by destruction of the hydrogen bonds maintaining the double-stranded state. *Compare* renaturation.

Desert–Land on which the vegetation is absent or sparse, is usually shrubby, and is characterized by an arid climate, hot to cool.

Desertification–Desertification means the degradation of land in dry areas of the world. It is not the spread of existing deserts, but rather the loss and destruction of healthy fertile soils. Loss of topsoil and soil fertility results in declining production of crops and livestock. Overall the problem is caused by people putting too much pressure on delicate soils and ecosytems. The main causes of degradation include:

Deforestation–Trees hold soil together and help water the land by channeling rainwater into the soil. When they are chopped down, the soil is eroded by the elements and is unable to hold water;

Overfarming–Overfarming the land eventually drains the soil its nutrients, leaving it unable to produce crops;

Overgrazing–Too many livestock, such as goats or cows, strip the soil of its vegetation and expose it to erosion by wind and water;

Poor irrigation practices–Bad irrigation can lead to waterlogging and salinisation of soil.

Over one-quarter of the Earth's land surface has suffered erosion and soil degradation.

Detassel–To remove, by cutting or pulling, the tassels that bear staminate flowers in order to prevent selfing during hybridseed production of corn.

Dethatch–The procedure of removing an excessive thatch accumulation either (i) mechanically, as by vertical mowing, or (ii) biologically, as by topdressing with soil.

Development–A process of economic and social transformation that defies simple definition. Though often viewed as a strictly economic process involving growth and diversification of a country's economy, development is a qualitative concept that entails complex social, cultural, and environmental changes. There are many models of what 'development' should look like and many different standards of what constitutes 'success'.

Developmental stage–Discrete portion of the life cycle of a plant, such as vegetative growth, reproduction, or senescence. Several published systems are in use for various crops to subdivide the broad stages. *Usage:* Preferred to growth stage (except when growth stage is part of the name of a published system).see *also* bloom, early; bloom, full; bloom, late; boot stage; grain maturity.

Diallelic–*Search in* balanced diallelic; unbalanced diallelic.

Differential centrifugation–A method of separating subcellular particles by centrifuging cell extracts at successively higher speeds. Separation is based on differences in sedimentation coefficients that are roughly proportional to particle size. Large particles (*e.g.*, nuclei, chloroplasts, or mitochondria) are sedimented at lower speeds than small particles (*e.g.*, ribosomes).

Differentiation–1. The process leading to differences in the amount, character, and location of cellular components, cells, tissues, and organs of an organism: specialization physically, chemically, physiologically, and morphologically. In particular, the maturation of a cell with unique traits (*e.g.*, a xylem vessel), different from the meristematic cells from which it arose and different from other cells of similar age that followed a different course. *Compare* development.

2. The process of biochemical and structural changes by which cells become specialized in form and function.

Digestible dry matter–Feed intake minus feces expressed as a per cent of feed dry matter consumed.

Digestible energy intake–Feed consumption expressed as units of digestible energy.

Digestible protein–Feed protein minus feces portion (nitrogen x 6.25) expressed as a percentage of amount in feed.

Diploid–(1) Characterizing a cell in which all chromosomes, except sex chromosomes, are two in number and are structurally identical within the limits of detection with those of the species from which it was derived. (2) Such a cell or organism.

Direct repeat–Two or more regions of DNA that have identical (or nearly identical) nucleotide sequences in the same orientation within one DNA molecule. Direct repeats may be either contiguous or separated on the same molecule. *Compare* inverted repeat.

Dirty seed–*Usage:* For tall fescueseed, endophyte-infected seed is the preferred term.

Divot–A small piece of turf severed from the soil by a golf club or the twisting-turning action of a cleated shoe.

DNA polymerase–One of several enzymes that synthesize a new DNA strand complementary to a template strand by adding nucleotides one at a time to a 3'-OH end.see *also* polymerase chain reaction.

DNase *Deoxyribonuclease*–Any enzyme that degrades DNA, including endonucleases and exonucleases.

Dormant seeding–Planting seed during late fall or early winter after temperatures become too low for seed germination to occur until the following spring.

Dormant turf–Turfs that have temporarily ceased shoot growth as a result of extended drought, heat, or cold stress.

Dry matter–The substance in a plant remaining after oven drying to a constant weight at a temperature slightly above the boiling point of water.

Dry matter disappearance–(1) Grazing: Forage present at the beginning of a grazing period plus growth during the period minus forage present at the end of the period. (2) Digestibility: Loss in dry weight of forage exposed to in vitro digestion.

Dry weight–Moisture-free weight.

Earthworm casts–*Search in* castings, earthworm.

Eckhardt method–A rapid screening method in electrophoretic analysis of plasmid DNA from recombinant colonies. The colonies may be lysed and the DNA applied directly to an agarose gel, or the colonies may be lysed in the well of the gel. Plasmids containing insert DNA are larger than the normal plasmid vector molecules, which are easily distinguished by differences in electrophoretic mobility. Series of biotypes within a species that shows a genetic gradient correlated with a gradual environmental gradient.

Ecological balance–Stability in an ecosystem achieved through the development of equilibrium among its various components. This does not imply that the community is static. It is subject to natural variations associated with ecological succession and other influences such as fire, disease and climate change, but the system is normally sufficiently elastic to make the necessary adjustments without major displacement of the balance. Human intervention that includes the introduction or removal of plants and animals, pollution of the environment and destruction of habitat is now a main cause of imbalance in many ecosystems.

Ecological niche–Role of an organism in an ecological system.

Ecological race–Group of local populations within a species in which individuals have similar environmental tolerances. Wide-ranging species may consist of many ecological races.

Ecology–Originally defined by Ernst Haeckel in 1866, ecology is the study of the relationships that develop among living organisms and between these organisms and the environment.

Ecotone–The transition zone / region that forms the boundary between two adjacent communities.

Efficiency–Degree to which a plant or vegetation converts radiant energy into organic compounds; efficiency may depend on degree of utilization of necessary environmental components.

Electroblotting–The electrophoretic transfer of macromolecules (DNA, RNA or protein) from a gel in which they have been separated onto a support matrix such as a nitrocellulose sheet. This procedure is an alternative to the capillary transfer used in techniques such as Southern and northern blotting.

Electrophoresis–A technique for separating different types of molecules based on their patterns of movement in an electrical field.

ELISA–Enzyme-linked immunosorbent assay: An immunological technique employing two antibodies. The primary antibody is specific for the antigen of interest, and the second is an antiglobin antibody to which an enzyme, such as horseradish peroxidase, has been attached. The primary antibody is first bound to the antigen and then the antiglobin antibody is bound to the primary antibody. Assays for the enzyme activity thus indicate the location and amount of primary antibody present and, indirectly, that of the antigen.

Embryo percent–(1) The amount of embryo compared with endosperm and otherseed parts. (2) The percent of embryo in the wholeseed.

Embryogenesis–The process of initiation and development of embryos from zygotes or, in somatic embryogenesis, of complete bipolar plant structures from somatic cells that parallel the developmental pathway of zygotic embryos. In the latter case, such embryos may be either adventitious or de novo in origin. *Synonym:* embryogeny.

Embryoid– Mass of cells or tissue that resembles an embryo; *i.e.*, an embryolike structure.

Enclosure–(1) An area enclosed by fence or wall to confine animals. (2) A caged or fenced area within a pasture to exclude grazing.

Endarch–Primary xylem in which the development is from the center towards the outside. Typical of stems. *Compare* exarch.

Endogen–A plant in whose primary stem the vascular bundles are not arranged in concentric circles and whose older stems do not have distinct wood and bark layers.

Endonuclease–An enzyme that cleaves within the polynucleotide chain.see *also* nuclease; restriction endonuclease.

Endophyte absence–Absence of the endophytic fungus *Acremonium coenophialum* in the fescue tissue, plant, or population being discussed.

Endophyte-freeseed–Tall fescueseed free of the *Acremonium coenophialum* endophyte.

Enhancer element–A DNA sequence found in eukaryotes and certain eukaryotic viruses that acts in *cis* in either orientation to increase the use of a promoter and transcription from a gene when located up to several kilobases from the gene it affects. *Synonym:* enhancer sequence

Environment–A combination of the various physical and biological elements that affect the life of an organism. Although it is common to refer to 'the' environment, there are in fact many environments eg, aquatic or terrestrial, microscopic to global, all capable of change in time and place, but all intimately linked and in combination constituting the whole earth/atmosphere system.

Environmentally-sound–The maintenance of a healthy environment and the protection of life-sustaining ecological processes. It is based on thorough knowledge and requires or will result in products, manufacturing processes, developments, etc. which are in harmony with essential ecological processes and human health.

Enzymatic degradation–Chemical breakdown of a given substance by the specific enzyme catalyst for that particular chemical reaction of a biological process.

Epigeal germination–A type of germination in which the cotyledons are raised above the ground by elongation of the hypocotyl.

Episome–A plasmid that can integrate into the bacterial chromosome.

Ergopeptine; ergopeptide–A class of alkaloids produced by clavicepitaceous fungal organisms, including ergovaline, ergotamine, and related ergotlike alkaloids.

Escape–A plant introduction that has accidentally escaped from cultivation.

Ether extract–Fats, waxes, oils, and similar plant components that are extracted with warm ether in chemical analysis.

Etiolated–Characterized (as a result of growth in the absence of light) by the development of a number of symptoms such as yellowing, elongation, thin stems, and failure of leaf expansion.

Euploid–(1) Characterizing a cell in which the nucleus contains exact multiples of the haploid number of chromosomes. (2) Such a cell or organism.

Exarch–Primary xylem in which the development is from the outside, toward the center of axis. Typical of roots. *Compare* endarch.

Exonuclease–An enzyme that cleaves nucleotides one at a time from either the 3' or 5' end of a polynucleotide chain.

Explant culture–The maintenance or growth of an explant in culture.

Explant tissue–The excised tissue of a plant (orseed) used to establish a cell or tissue culture system.

Extrachromosomal–Located elsewhere than the nuclear chromosome, such as in an organelle or on a plasmid.

Extrusion cooking–A process where a mash or granular mixture of starch and/or proteins of cereals or oilsearch Inds is preconditioned by exposure to steam and then cooked in a continuous extruder. The extruder consists of a series of spirals that force the cereal mash into an increasingly restricted area, whereby the temperature and pressure are increased rapidly, reaching 300 to 400 psi (~2-2.8 MPa) and 300 to 500 °F (~150-260 °C) for a short period of time (15-20 s). The pressure is released and the hot, plasticized material expands as it passes through a series of dyes. The extruded material is then dried. The process is used in production of textured vegetable proteins, pet foods, instant foods, and many other products.

Facilitated recurrent selection–A type of recurrent selection in which genetic male sterility is maintained in the population to maintain heterozygosity and genetic diversity and to permit the recombination and shifting of gene frequencies.

Fallow–Land left unsown for a season, giving it a rest from cropping.

Fat-free body weight–*Search in* body weight, fat-free.

Feed forward–The effect of process inputs (the levels of factors entering into a process) on the process rate. In positive feed forward, the inputs amplify the rate; in negative feed forward, they diminish the rate.

Feeding value–(1) Characteristics that make feed valuable to animals as a source of nutrients. (2) The combination of chemical, biochemical, physical, and organoleptic characteristics of forage that determine its potentials to produce animal meat, milk, wool, or work. Considered by some as synonymous with nutritive value.

Fermentation–Large-scale culture of cells suitable for recovery of cell products, including various chemical or pharmaceutical compounds, or biomass *e.g.* alcohol are produced by this process.

Fertigation–The application of fertilizer through an irrigation system.

Fertilizer burn–*Search in* foliar burn.

Fescue foot–Red and swollen skin at junction of hoof of animals grazing tall fescues, followed in advanced stages by gangrenous sloughing off of hoofs, tail tips, and ear tips, along with loss of appetite and emaciation. Problem is most severe in cold, rainy, or overcast weather.

Fescue toxicity–(1) Exposure of an animal to the endophyte-infected tall fescue plant, eliciting a toxic response in the animal. (2) The state of being toxic to the animal.

Fiber–(1) A long, thick-walled cell (*e.g.*, in cotton) serving to strengthen an organ. (2) A unit of matter characterized by a length at least 100 times its diameter or width.

Fiber fineness–A relative measure of size, diameter, linear density, or mass per unit length of fiber, expressed in a variety of units.

Fiber length distribution–A graphic or tabular presentation of the proportion or percentage of fibers having different lengths.

Fiber, textile–A generic term for the various types of matter that form the basic elements of textile fabrics and other textile structures.

Fibrograph test beard–The portion of a test specimen (*e.g.*, of cotton) that has been combed into a beardlike shape and that protrudes from the outside of a pair of fibrograph combs.

Field burning–Burning plant residue after harvest to (i) aid in insect, disease, and weed control; (ii) reduce cultivation problems; and (iii) stimulate subsequent regrowth and tillering of perennial crops.

Fines–Material that passes through a screen whose openings are smaller than the specified minimum size of the product being processed.

Fistula–A surgical opening, duct, or passage from a cavity or hollow organ of the body.

Flail mower–A mower that cuts turf by impact of blades rotating in a vertical cutting plane relative to the turf surface.

Flavor–The simultaneous physiological and psychological response obtained from a substance in the mouth that includes the senses of taste (sweet, sour, salty, bitter), smell (fruity, pungent), and feel. The sense of feel as related to flavor encompasses only the effect of chemical action on the mouth membranes, such as heat from pepper, coolness from peppermint, and the like. No reliable correlation of taste with chemical structure has yet been possible. Flavor is a critical factor in the acceptability of foods, medicines, confectionery, and beverages.

Flint–Maize variety with very hard kernels.

Flowering stage–The physiological stage of a grass plant in which anthesis (blooming) occurs, or flowers are visible in nongrass plants.

Flush–Growth that is produced during a short period.

Fodder–1. Coarse grasses such as corn and sorghum harvested with the seed and leaves green or alive, then cured and fed in their entirety as forage.

2. Animal food which has been grown or collected specifically for livestock. For example, hay, straw, feed grain, grass, kale.

Fold-back–The double-stranded structure formed by intrastrand reassociation between inverted repeat sequences present in the same DNA strand.

Foliar burn–Injury to shoot tissue caused by dehydration due to contact with high concentrations of chemicals (*e.g.*, certain fertilizers and pesticides).

Food deficit–A shortage of foodstuffs in relation to the recommended food needed.

Food security–Access to sufficient nutritious food at all times.

Food surplus–A surplus of foodstuffs in relation to the recommended food needed.

Foot printing, wilt–Temporary foot impressions left on turf when flaccid leaves of grass plants suffer incipient wilt and have insufficient turgor to spring back after being stepped on.

Forage allowance–The relationship between the weight of forage dry matter per unit area and the number of animal units or forage intake units at a given time; a forage-to animal relationship. The inverse of grazing pressure. *Usage:* For a more specific term, substitute herbage or browse for forage.

Forage intake unit–An animal with a rate of forage consumption equal to 8 kg dry matter per day. *Usage:* Assuming that one animal unit has a dry matter intake rate of 8 kg per day, any animal may be represented as a certain fraction or multiple of the animal unit, based solely on its rate of forage intake per day. An animal that has a forage intake rate larger or smaller than 8 kg dry matter per day will have an animal-unit equivalent that is a proportionate fraction or multiple of one animal unit. In a publication, the term *forage intake unit* should be followed by a description in a standard format, including at least the following: forage species and cultivar, stage of growth, plant height, and forage mass.

Forb–Any herbaceous broadleaf plant that is not a grass and is not grasslike.see *also* legume.

Forest land–Land on which the vegetation is dominated by forest or, if trees are lacking, the land bears evidence of former forest and has not been converted to other vegetation.

Form–An infraspecifc taxon in botanical nomenclature, below the rank of variety. *Abbr.*: f. *Usage:* The abbreviation in roman type; the name, in italics; no capitals.

Forward creep–A method of creep grazing in which dams and offspring rotate through a series of paddocks with offspring as first grazers and dams as last grazers. A specific form of first-last grazing.

Foundation seed–Foundation seed is the progeny of breeder, select, or foundation seed handled to maintain specific genetic purity and identity. Production must be acceptable to the certifying agency.

Fpasture, rotation–Pastureland used for a few seasons and then plowed for other crops.

Fractionation–Separation into components, as by distillation, crystallization, or physical separation.

Frame-shift mutation–Mutation involving a change in the reading frame.

Frequency of clip–Distance of forward travel between successive cuts of mower blades.

Frontal grazing–A grazing method that allocated forage within a land area by means of a sliding fence that livestock can advance to gain access to ungrazed forage.

Full (rumen content)–Amount of ingested feed or water present in the rumen.

Fungus–*Usage:* Too vague to be an acceptable term for the fungus *Acremonium coenophialum* living symbiotically in tall fescue. Similarly, 'fungal absence', 'fungal incidence', 'fungal presence', 'fungus-free', and 'fungus-infected' are too vague without the context clearly specified.

Fusion–Joining of the membrane of two cells, thus creating a daughter cell that contains the nuclear material from parent cells.see *also* protoplast fusion.

GA–*See* gibberellins.

Gametocide–An agent used to selectively kill either male or female gametes.

Gap–A missing section on one of the strands of a DNA duplex. The DNA will therefore have a single-stranded region.

Gel filtration–A type of fractionation procedure in which molecules are separated from each other according to differences in size and shape; the action is similar to that of molecular sieves. Dextran gels (three-dimensional networks of polysaccharide chains) are usually used in this method.

Gel–The inert matrix used for the electrophoretic separation of nucleic acids or proteins.see *also* electrophoresis; polyacrylamide gel electrophoresis; sequencing gel.

Gene bank–The collection of cloned DNA fragments that, ideally, represent all the sequences of a single genome. *Synonyms:* gene library; genomic library; library.

Gene mapping–Determination of the relative locations of genes on a chromosome or within a genome.

Gene sequencing–Determination of the identity and order of nucleotide bases in a strand of DNA.

Generation–Symbols for generations include BC (backcross), F (filial, or hybrid), M (mutant), R (regenerant), S (selfed), and Syn (synthetic). The generation number is usually indicated by a subscript: F_2, R_0, S_1, etc., but Syn 2. *Compare* C for cycle of selection (not necessarily a generation; no subscript).

Genetic correlation–Correlation between the genotypic values of two characters, or of two individuals with respect to the genetic character or different characters.

Genetic engineering–1. Inserting genes from one source into another using molecular techniques.

2. The use of in vitro techniques to produce DNA molecules containing novel combinations of genes or other sequences in living cells that make them capable of producing new substances or performing new functions. *Usage:* A popular term for such technologies as a whole.

Genetic shift–A change in the germplasm balance of a cross-pollinated variety, usually caused by environmental selection pressures.

Genetically modified organism (GMO)–The product of genetic engineering.

Genomic library–*Search in* gene bank.

Geographic range–Geographic limits of the ecologic range; geographic extent of actual occurrences of a species.

Germination (1) Resumption of active growth by theseed embryo, culminating in the development of a young plant. (2) Inseed laboratory practice: emergence and development from theseed embryo of those essential structures, which, for the kind ofseed in question, are indicative of the ability to produce a normal plant under favorable conditions.

Gibberellic acid–One of the gibberellins. *Abbr.*: GA_3.

Ginned lint–Cotton fibers after they have been removed from theseed.

Glutenin–The dilute alkali-soluble fraction of proteins in wheat.

GM foods–Foodstuffs that have had their genes changed (genetically modified–GM) in order to improve their productivity.

Grade, cotton–A system of scales used in classifying cotton with regard to color, trash content, and preparation.

Grain grade–Market standard established to describe the amount of contamination, grain damage, immaturity, test weight, and marketable traits.

Grain, percent–Threshed grain weight x 100/(threshed grain weight + stalk weight).

Grain-to-stalk ratio–Threshed grain weight/stalk weight.

Grass–Member of the plant family Poaceae.

Grassland–(1) Land on which the vegetation is dominated by grasses. *Compare* pastureland; rangeland. (2) More generally, any plant community in which grasses and/or legumes make up the dominant vegetation.

Grazable forest land–Forest land that produces, at least periodically, sufficient understory vegetation that can be grazed. Forage is indigenous or, if introduced, is managed as though it were indigenous. *Synonyms:* grazable woodland; woodland range; forest range.

Grazer–(1) Animal that grazes in situ grass as herbage. (2) On experimental pastures, animals that may or may not remain on specified pasture treatment for the entire grazing period or season but are of a kind or physiological condition not necessarily represented on all pasture treatments for the entire grazing period or season.

Grazing event–The length of time that an animal grazes without stopping. *Compare* grazing period.

Grazing land management–The manipulation of the soil-plant-animal complex of the grazing land in pursuit of a desired result. *Usage:* For more specific applications, substitute the appropriate term (*e.g.*, grassland) for grazing land.

Grazing management unit–The grazing land area used to support a group of grazing animals for a grazing season. It may be a single area, or it may have a number of subdivisions.

Grazing period–The length of time that grazing livestock or wildlife occupy a specific land area.

Grazing pressure–The relationship between the number of animal units or forage intake units and the weight of forage dry matter per unit area at a given time; an animal-to-forage relationship. The inverse of forage allowance.

Grazing system–A defined, integrated combination of animal, plant, soil, and other environmental components and the grazing method or methods by which the system is managed to achieve specific results or goals. *Usage:* Descriptive common names may be used; however, the first usage of a grazing system name in a publication should be followed by a description using a standard format. This format should include at least the following information: number, size, kind, slope, erosion status, and soil classification of land units; number, kind, sex, size, and age of livestock; duration of use and nonuse periods for each unit in the system; grazing methods; types of forage; geographic location and elevation; and type of climate and mean annual and seasonal temperatures and precipitation.

Green Revolution–The term given to the rapid increase in crop production brought about in the late 1950s and 1960s by a combination of increased fertiliser use and the introduction of new high yielding varieties of grain.

Groat–The caryopsis or kernel of oat after the husk has been removed.

Gross energy–The amount of heat that is released when a substance is completely oxidized in a bomb calorimeter containing oxygen at 25 to 30 atmospheres pressure (~2.5-3.0 MPa).

Growth analysis–Mathematical analysis of crop or plant growth (following original concepts of Blackman and of West, Briggs, and Kidd) using relative growth rate (RGR), net assimilation rate (NAR), leaf area growth rate (LAR), and crop growth rate (CGR). Generally involves the relationship of these variables to other crop or environmental variables as integrated over time (sampling interval).

Growth rate–A rate of change in growth (*e.g.*, in number size or mass) of an organ, organism or community. (1) Absolute growth rate is represented by dx/dt as change in amount per unit time. (2) Specific (or relative) growth rate is dx/xdt, change per unit amount per unit time (with units of 1/time). (3) Crop growth rate is a specific plant growth analysis term denoting the absolute growth rate of mass per unit land, A, thus dw/Adt, usually in units of grams per meter per day ($g\ m^{-1}\ d^{-1}$). *Usage:* Relative growth rate (RGR) is the usual term in plant growth analysis, whereas specific growth rate is used in most other disciplines.

Growth stage–*Search in* developmental stage.

Guide sequence–An RNA molecule (or a part of one) that hybridizes to eukaryotic mRNA and aids in the splicing of intron sequences from a primary transcript. Guide sequences may be either external or internal to the mRNA being processed and may hybridize either to the intron or exon sequences close to the splice junction.

Habituation-The ability of cells to grow in the absence of normally required and exogenously supplied growth factors as a result of shifts in metabolism, especially as in the case where cultured cells become autonomous (habituated) for the production of phytohormones. *Usage:* Habituated is the preferred term.

Hardseed–Search ind that is dormant due to a seed coat impervious to either water or oxygen.

Hardiness–Capability of an organism to withstand environmental stress. *Synonym:* stress tolerance.

Head components–Components of the inflorescence of grains and grass crops; generally, the grain rather than vegetative structures, but may include individual vegetative structures such as rachis, peduncle, and pedicel.

Heading–The developmental stage of a grass plant from initial emergence of the inflorescence from the boot until the inflorescence is fully exserted.

Heat unit–*see* heat sum.

Hemicellulose–Polysaccharides that accompany cellulose and lignin in the cell walls of green plants; differs from cellulose in that it is soluble in alkali and with acid hydrolysis gives rise to uronic acid, xylose, galactose, and other carbohydrates as well as glucose.

Herbage–The biomass of herbaceous plants, other than separated grain, generally above ground but including edible roots and tubers.

Herbicides–Chemicals used to kill plants and inhibit their growth, chemically related to hormone.

Heteroduplex–A DNA duplex formed by hybridization of complementary regions of single strands derived from different parental double-stranded DNAs. If the single-strand sequences are not completely complementary, then single-stranded regions will be present in the heteroduplex molecule and can be visualized and mapped using electron microscopy.

Heterogeneous–Characterizing individuals with nonhomozygous allelic genetic constitutions, or populations of plants with individual plants of different genetic constitution. *Usage:* Not applied to a single locus, where the term is heterozygous. *n.* heterogeneity.

Heteroploid–*Of a cell culture*: comprised of cells possessing nuclei that contain chromosome numbers other than the diploid number. Thus, a heteroploid culture is one containing aneuploid cells. *Usage:* Not used to describe individual cells.see *also* mixoploid;–ploid.

High-intensity grazing–*Usage:* Not an acceptable term (it is a relative concept, best described in terms of management and method).see grazing management; intensive grazing management.

High-velocity microprojectile transformation–A method for introduction of DNA into plant cells. Tungsten or gold particles are coated with DNA and propelled at high speed through the target cell walls by means of an electrical or gunpowder discharge.

Histochemistry–The chemistry of cells and tissues.

Homogenotization–The replacement of one copy of a DNA sequence (usually a gene) with an altered copy by genetic and recombinant DNA techniques. The target gene is cloned and modified in some way, such as by inserting a transposable element containing a marker gene into the target gene. Introduction of the modified gene back into the target host cell may allow the modified gene to be exchanged for the target gene by in vivo genetic recombination. Selection for expression of the marker gene facilitates identification of modified cells.

Homology–(1) The extent to which two nucleic acid molecules have the same nucleotide sequence or two proteins have the same amino acid sequence. Homology can be determined by direct comparison of sequence data, or estimated by DNA-DNA or DNA-RNA hybridization. (2) The degree of identity between chromosomes or chromosome segments.

Host–1. A cell or organism used for supporting the replication of a plasmid, virus, or other form of foreign DNA, or for the production of cloned substances.

2. Organism infested by a parasite or pathogen.

Host-vector system–Combination of DNA-receiving cells (host) and DNA-transporting substance (vector) used for introducing foreign DNA into a cell.

Hot-water insoluble nitrogen–Fertilizer nitrogen not soluble in boiling-point hot water (100 °C; 212 °F); used to determine activity index of urea forms. *see* nitrogen activity index.

Hybrid released translation–A technique to identify the protein encoded by a cloned DNA sequence, usually a cDNA. The cloned DNA is bound to a filter and hybridized to a mixture of mRNA. Only the hybridizing mRNA is used in subsequent in vitro translation reactions for analysis of the polypeptides formed.

Hybrid variety–First-generation (F_1) progenies from a cross, produced through controlling the pollination, between (i) two inbred lines; (ii) two single crosses; (iii) a single cross and an inbred line; (iv) an inbred line or a single cross and an open-pollinated or a synthetic variety; or (v) two selected clones,seed lines, varieties, or species.

Hybridization–(1) The pairing of complementary DNA or RNA strands to give stable DNA-DNA or DNA-RNA duplexes. The efficiency of hybridization is a test of sequence homology. (2) Production of offspring, or hybrids, from genetically dissimilar parents. The process can be used to produce hybrid plants (by cross-breeding two different varieties) or hybridomas (hybrid cells formed by fusing two unlike cells, used in producing monoclonal antibodies).

Hydrocyanic acid (HCN)–A poison produced as a glucoside by several plant species, especially sorghums. *Synonym:* prussic acid.

Hydrosearch inding–Planting seed in a water mixture by pumping through a nozzle that sprays the mixture onto aseed bed. The water mixture may also contain addends such as fertilizer and mulches.

Hydroxyapatite [$Ca_5(PO_4)_3OH$]–A form of calcium phosphate that can bind nucleic acids. Under certain conditions it will bind double-stranded but not single-stranded DNA. It can therefore be used to fractionate DNA preparations and to determine the extent of hybridization of two classes of single-stranded molecules. *Synonym:* hydroxylapatite. *Abbr.*: HAP.

Hypogeal germination–A type of germination in which the cotyledons remain below the ground while the epicotyl grows and emerges above the ground.

Hysteresis–A phenomenon evidenced by the fact that at a given relative humidity,seeds or grain may reach two different equilibrium moisture contents: one by increasing the relative humidity from a low level, and another by decreasing the relative humidity from a high level.

IAA, ***Indoleacetic acid***–a natural hormone.see *also* auxin.

Illegitimate recombination–Recombination between DNA species that show little or no homology. Illegitimate recombination is promoted by transposable elements.

Immune–Not subject to attack by a specified pest. Immunity is absolute, *i.e.* power of resistance for various kind of attacks.

IMP–Inosine 5'-monophosphate.

***in vitro* dry matter disappearance**–A gravimetric measurement of the amount of dry matter lost upon filtration following the incubation of forage in test tubes with rumen microflora, usually expressed as a percentage: (Wt. dry matter sample– wt. residue) / (weight dry matter sample). *Abbr.*: IVDMD. *Synonym:* in vitro digestible dry matter (IVDDM).

Usage: The term disappearance is preferred to digestibility. *Compare:* in vitro organic matter disappearance.

***in vitro* propagation**–Propagation of plants in a controlled, artificial environment using plastic or glass vessels, aseptic techniques, and a defined growth medium.

***in vitro* translation**–The synthesis of proteins in the test tube using purified mRNA molecules and cell extracts containing ribosomal subunits, the necessary protein factors, tRNA molecules, and aminoacyl tRNA synthetases. ATP, GTP, amino acids, and an enzyme system for regenerating the nucleoside triphosphates are added to the mix. Prokaryotic translation systems are usually prepared from *Escherichia coli* or *Bacillus stearothermophilus*. Eukaryotic systems usually employ rabbit reticulocyte lysates or wheatgerm extracts. *Synonym:* cell-free translation.

***in vivo* nylon bag technique**–System of determining dry matter disappearance of forage placed in fine-mesh nylon bags, either placed in the rumen or suspended in the rumen from a canula cover of a fistulated animal.

Indehiscent–Not splitting open at maturity.

Indeterminate inflorescence–Flowers arising laterally and successively as the floral axis elongates.

Indigenous–A plant native to the location where it is growing.

Inducible–The capability of a gene (or its promoter) to increase transcription when exposed to an inducing agent.

Initiation–(1) Inception of a new organ, as in organ formation at a meristem (*e.g.*, leaf and floral part initiation at the shoot apical meristem), activation of an existing meristem (cambia in shoots and potential underground storage organs), or development of a new meristem (lateral root initiation, adventitious bud initiation). (2) A morphological change of the growing point from the vegetative condition to the floral condition.

Inoculation–The physical process of applying inoculant to seed or soil.

Insecticides–Chemicals used to kill insects, especially those considered undesirable by farmers.

Insert–The piece of foreign DNA introduced into a vector molecule.

Insertion sequence–A class of small (several hundred base pairs) bacterial DNA elements capable of transposition and thus inactivating a gene into which it inserts. Its presence is generally detected by loss of the target gene's function. *Abbr.:* IS.

Integration–The recombinational insertion of a small DNA molecule such as a virus into a larger one such as its host. Circular molecules require one crossover for insertion; linear molecules require two.

Intermittent grazing–A method that imposes grazing for indefinite periods at irregular intervals.

Intersearch inding–Search inding between sod plugs, sod strips, rows, or sprigs, orseeding into turf to improve the stand or alter its composition.

Introduced species–A species not part of the original fauna or flora of the area in question; *i.e.*, brought by human activity from another geographical region.

Inverse nitrogen–Yield concept proped by Wilcox; according to him, the yield of a crop is inversely proportional to its nitrogen content *i.e.*, Y = k/n; where y is the yield, n is the percentage of nitrogen in the crop, k is constant (the value of k as found by Wilcox is 318 lb/acre).

Inverted repeat–Two or more regions of DNA that have identical (or nearly identical) nucleotide sequences in the opposite orientation within a single DNA molecule. Inverted repeats may be either contiguous or separated on the same molecule.

Irrigation, automatic–Hydraulic-electric control of irrigation in response to a transducer that senses plant needs. The term is commonly used more loosely to refer to hydraulic or electrically actuated valves manually present on a time-based controller.

Irrigation, semiautomatic–An irrigation system in which valves respond directly to a manually operated remote-control switch.

Irrigation–The provision of water for crops in areas where the natural precipitation is considered inadequate for crop growth.

Isochromosome–Chromosome with arms equal in length and genetically identical, but with the loci in reverse order; produced when the centromere splits transversely.

Isogenic–Differing in genotype by only one or a very few genes.

Leguminous plants–A large group of pod-bearing plants (including acacia, peas, beans, alfalfa and clover) whose roots contain nodules with nitrogen-fixing bacteria, and are thus able to absorb nitrogen from the atmosphere and converted into a form useful to plants.

Meteorology–Study on the atmosphere and interpretation of atmospheric processes.

Micrometeorology–Study of variations in meteorologic conditions overy very small areas, such as hillsides, forests, drainage area, rivers, individual experimental plots etc.

Minimum tillage–Soil management system in which the crop is seeded directly into the field without ploughing or discing.

National conservation strategies–Plans that highlight country-level environmental priorities and opportunities for sustainable management of natural resources, following the example of the World Conservation Strategy published by the World Conservation Union (IUCN) in 1980. Though governments may support preparation for the strategies, they are not bound to follow IUCN's recommendations.

Organic farming–Production system that avoids or largely excludes the use of synthetically produced fertilizers, pesticides, growth regulators, and livestock feed additives. As far as possible, it relies on crop rotations, crop residues, animal manures, legumes, green manures, off-farm organic wastes, and aspects of biological pest control to maintain soil productivity and tillage, to supply plant nutrients, and to control insects, weeds, and other pests.

Organochlorides–A group of organic compounds that contain chlorine (Cl). They have a variety of forms and uses including aerosol propellants, plasticisers, transformer coolants (PCBs) and food packaging (PCVs), but their greatest use was as pesticides, in the form of DDT, Aldrine and Lindane. However, with time many pests have developed immunity to them and it has also become clear that the characteristics that made them good pesticides–persistence, mobility and high biological activity–also posed dangers for the environment. Organochlorides accumulate in the fatty tissue of animals, and through biomagnification in the food chain may reach toxic levels in predators. Because of side effects such as sterility, birth defects, cancer and damage to the nervous system, they have been banned or had their use severely restricted in most parts of the world.

Organophosphoruus compounds–A group of pesticides that work by blocking the central nervous systems of the organisms exposed to them. Malathion and diazonon are the most commonly used organophosphates. They are highly effective against insects, but break down rapidly in the environment and do not bioaccumulate. For these reasons, they are preferred over organochloride pesticides. Although generally considered safer than the organochlorides, they are highly toxic to humans and other mammals and may be carcinogenic.

Pastoral farming–A form of agriculture based on the herding of grazing animals such as cattle, sheep, goats and camels, common in the world's arid regions.

Pesticides–Chemical products designed to kill or restrict the development of pests. They include fungicides, herbicides and insecticides. Pesticides range from relatively simple elements such as sulphur (S) to complex chemical compounds such as chlorinated hydrocarbons and may be broad-spectrum or narrow-spectrum agents. Pesticides also vary in their persistence in the environment, and in general, the longer they remain chemically stable the greater is their potential for environmental damage. The use of pesticides has undoubtedly benefited society, by preventing disease and improving the food supply. At the same time, ignorance of the environmental impact of pesticides, the indiscriminate use of certain products and inadequate control of the production and use of pesticides has created problems for wildlife and natural vegetation and has threatened human health.

Pesticides–Chemical products designed to kill or restrict the development of pests, pests being organisms considered undesirable by those using them.

Proper resource pricing–The pricing of natural resources at levels which reflect their combined economic and environmental values.

Salinisation–The build up of salts in soil as a result of the capillary flow of saline water towards the surface. A common problem in areas where the land is irrigated.

Shifting cultivation–A system of cultivation common in the tropics, whereby forests and grassland are cleared, usually by fire, which provides ash to fertilise the land. After a few years, when the soil is exhausted, the cultivators abandon the plot and move elsewhere.

Soil erosion–The removal of top soil by water, wind and gravity. A natural process which can be hastened by human activity.

Subsidy–A financial benefit or form of assistance given to producers (*e.g.*, grants, loans, tax allowances) which enables them to sell or export goods at less than their costs of production, thus creating unfair competition.

Subsistence farming–The production of sufficient food and other necessities to meet the requirements of a farm unit, leaving no surplus for sale and little for storage.

Source: Many of these definitions have been taken from, or abridged from The Environment Dictionary *by David D Kemp (Routledge, London, 1998).*

Sustainable development–Sustainable development has as many definitions as subscribers. In essence, it refers to economic development that meets the needs of all without leaving future generations with fewer natural resources than those we enjoy today. It is widely accepted that achieving sustainable development requires balance between three dimensions of complementary change:

Economic (towards sustainable patterns of production and consumption)

Ecological (towards maintenance and restoration of healthy ecosystems)

Social (towards poverty eradication and sustainable livelihoods)

World Commission on Environment and Development–Established by the United Nations General Assembly in 1983 to examine international and global environmental problems and to propose strategies for sustainable development. Chaired by Norwegian Prime Minister Grò Harlem Brundtland, the independent commission held meetings and public hearing around the world and submitted a report on its inquiry to the General Assembly in 1987.

World Food Summit (WFS)–World Food Summit, held in 1996, at which governments pledged to halve the number of hungry poeple by 2015.

World Summit on Sustainable Development (WSSD): The World Summit on Sustainable Development takes place from 26 August–

04 September 2002 in Johannesburg, South Africa. Governments, UN agencies, and civil society organisations will come together to assess progress since the UN Conference on Environment and Development held in Rio in 1992 (hence the title 'Rio + 10' for the Johannesburg meeting). Sustainable development is defined in the report from the Rio meeting as being 'economic progress which meets all of our needs without leaving future generations with fewer resources than those we enjoy'.

Chapter 13
Rose Glossary

Alba Roses–The third ancient group, hybrids between a rose of the *Rosa canina* section such as *Rosa mollis* and *Rosa gallica*, or possibly a Damask. Alba roses are always white or pale pink flowered, sweet scented, with bluish leaves. They make large bushes up to 2.5 m, with rather few thorns.

Antique Roses–*Search* in Old Roses.

Bourbon Roses–Before the opening of the Suez Canal (1871) the Ile de Bourbon, now Reunion, east of Madagascar, was an important stopping point for French ships sailing between the Far East and Europe. By 1817, Rosa chips had been grown there, as had the Autumn Damask from Europe, and they were apparently planted together as mixed hedges. A French botanist, M. Breon, noticed an intermediate between the two types in such a hedge and moved it to his botanic garden whence seeds were sent to M. Jacques in Paris. He raised from them the original Bourbon rose, which was painted by Redout in 1824. It had semi-double, bright pink flowers, good scent inherited from the Damask, and good autumn flowering. The Bourbon roses were bred by crossing this original Bourbon with Gallica and Damask hybrids.

Boursault Roses–These large shrubs or semi-climbing roses raised in the early nineteenth century, were said to be derived from *Rosa pendulina* x *Rosa chinensis*, but their chromosome number suggests that either *Rosa majalis* or *Rosa blanda* were used, not *Rosa pendulina* propeRosa *Rosa blanda* has also been used to breed thorn less roses in the USA.

Canina Roses–*Search* in Old Roses.

Centifolia Roses–(also called Cabbage Rose, Holland Rose, Rose des Peintres, or Provence Rose). The original Centifolia roses probably appeared around the end of the sixteenth century, from a cross between the Autumn Damask and an Alba. They make large, rather floppy bushes with very double flowers hanging on weak branches. A single or semi-double sport arose in the early nineteenth century, before which the group was sterile and the early varieties grown were all sports from the original cross.

China Roses–China roses had been cultivated and new varieties produced on a large scale in China for many centuries before the first were brought to gardens in Europe in 1792.

The dwarf, perpetual-flowering Chinas were mutants of the climber *Rosa chinensis*, some possibly hybridized with *Rosa odorata*. This mutation has been observed in cultivation with Little White Pet being a dwarf perpetual-flowering sport of Felicite et Perpetue. The original introductions from China were Slaters Crimson, China introduced in about 1792, and Parsons Pink China, now called Old Blush, in 1793. The original Chinas are dwarf, Up to 2 m, but usually around 1 m, and rather tender with red or pink single or loosely double flowers.

Climbing Roses (*Search* in Ramblers)–Many roses are natural climbers, and several of those *e.g. Rosa gigantea, Rosa chinensis* and *Rosa moschata* are important ancestors of modern roses. Furthermore, it has been noted that the gene for climbing is dominant over the dwarf (or bush) gene. Therefore it is not surprising to find many climbing Hybrid Teas and Floribundas; some of them arose as seedlings, others as sports of normal roses, such as Climbing Iceberg. Many of these climbing sports are summer flowering only, others have some later flowers. More recent large-flowered climbers, many raised by McGredy, are regularly repeat flowering.

Damask Roses–There are two groups of Damasks, the Summer Damask, once-flowering, and the Autumn Damask which has a second flowering in the autumn. Both have been grown since ancient times. The Summer Damask is a hybrid between *Rosa gallica* and *Rosa phoenicea*, a native of the eastern Mediterranean which looks like *Rosa multiflora*, with hairy leaves. The Autumn Damask is a hybrid between *Rosa gailica* and *Rosa moschata*. They are rather less hardy than the Gallicas, often taller, up to 2.5 m, with usually richly scented, red, pink or white flowers in loose clusters.

Damascena Roses–*Search* in Old Roses.

Eglanteria (Rubiginosa) Roses–*Search* in Old Roses.

English Roses: (abbrev. ER, DA)–This new group of roses, often called David Austin Roses, was introduced in 1969 by David Austin of England. These roses are an attempt to combine the best traits of both Old Roses and Modern Roses. David Austin has attempted to produce roses with the classic flower forms and fragrance of the Old Roses on plants that repeat bloom like the Modern Roses. Some of the popular English Roses are Abraham Darby, Graham Thomas, Heritage, and Mary Rose.

Filipes Roses–*Search* in Old Roses.

Floribunda Roses (abbrev. FB or FL) (Search in Modern Roses)–Floribundas were created about 1909 by crossing the Polyanthas with Hybrid Teas. They produce flowers in clusters, not singly like the Hybrid Teas. Floribundas are usually shorter plants than Hybrid Teas and tend to produce more flowers and smaller flowers than Hybrid Teas on shorter stems. Although Hybrid Teas provide excellent cut flowers, Floribundas are well suited as good landscape plants providing lots of coloRosa Many Floribundas are not very fragrant.

Foetida Roses–*Search* in Old Roses.

Gallica Roses–(Also called Rose of Provins) Grown since ancient times, these are varieties of *Rosa gallica*, a short suckering rose, native of southern Europe from France eastwards to central Turkey. They are very hardy, once-flowering, with strongly scented flowers ranging from pink, through crimson, to purple, height up to 1.5 m.

Grandiflora Roses (Search in Modern Roses)–The Grandiflora is a "manufactured" class—the class was invented for the rose Queen Elizabeth, introduced in 1954 by

Germain's Nursery in the USA. This rose was a cross of Charlotte Armstrong, a Hybrid Tea, and Floradora, a Floribunda. This rose is representative of the attempts at that time to produce a "different" rose (a mere 100 years after the first Hybrid Tea appeared) that would have the characteristic long stems, large beautiful blooms and pointed buds of the Hybrid Teas with the hardiness and flower clusters of the shrubbier Floribundas.

Grandifloras have a tendency to grow quite tall and produce full, large flowers. They come one to a stem as well as in clusters. The gangly growth habit is remniscent of their Tea heritage. The individual florets are larger than the standrad for Floribundas yet not usually as large as the huge blooms of the Hybrid Teas.

The United States recognizes this type of rose as a separate class in rose competitions while the International rose community lumps them in with the Hybrid Teas and often refer to the whole bunch of them as "large-flowered modern roses".

Grandifloras: Shining Hour, Queen Elizabeth, Sundowner, Prima Donna, John S. Armstrong, Lady Luck, Tournament of Roses, Gold Medal, Camelot, Ole, Sonia, Love.

Ground Cover Roses–The late 1980's witnessed the advent of genuine low-growing ground-cover Roses, specialty bred to spread without gaining height. Incidentally, these make wonderful hanging basket displays. Also beginning to make their presence felt are Climbing miniatures and Climbing Patioes, which will no doubt prove dually populaRosa

Hybrid Musk Roses–These roses were mostly raised by the Reverend Joseph Pemberton, at Havering-atte-Bower in Essex, using the rose Trier, a hybrid between a hybrid tea and a seedling of Aglaia, which had the musk rose in its ancestry. Thus the name Hybrid musk is rather far-fetched though generally accepted. Pemberton crossed Trier with various Hybrid Teas, producing several very beautiful, though sterile, seedlings; tall, repeat-flowering, scented shrubs, with large clusters of small flowers. Penelope(1923) and Felicia (1928) are well-known examples.

Hybrid Perpetual Roses–Over a thousand varieties of Hybrid Perpetual were raised in the latter part of the nineteenth century, of which probably less than a hundred survive today. The ancestor of this class was the Portland x China hybrid, Rose du Roi, which appeared in 1816. When crossed with both Hybrid Chinas (Gailica x China crosses) and Bourbons, in about 1835, a new class of Hybrides remontants or Hybrid Perpetuals was born. Some of the first were raised by Laffay, notably La Reine and Gloire des Rosomanes, the latest such as Arilaga in the early twentieth century.

The Hybrid Perpetuals are usually rather coarse growing, usually red, mauve, pink or white, often with huge flowers and strong shoots which need to be pegged down so that they produce flowers along their length. Their Portland ancestry makes these hardier than the Noisettes, but they still need protection in cold areas.

Hybrid Tea Roses–Hybrid Teas are easily the most popular class of roses today. Hybrid Teas as a group have large flowers with a high-pointed bud. They are excellent repeat bloomers, often blooming almost continually. They bloom one flower per stem on long sturdy stems making them excellent for cutting. Hybrid Teas come in a large variety of colors. Hybrid Teas are upright shrubs.

The rose "La France", bred in 1867, is classified as the first Hybrid Tea rose.

Kordesii Roses–A hybrid between *Rosa rugosa* and *Rosa wichuraiana* named Max Graf (I 9 I 9) was a sterile diploid, but Kordes succeeded in raising three seedlings from it in 1940 and 1950-1 which were fertile tetraploids, the beginning of a new race of Kordesii hybrids, such as Parkdirektor Riggers.

Macrantha Roses–Using *Rosa macrantha* Daisy Hill, possibly a hybrid between *Rosa canina* and *Rosa alba*, Kordes produced several hybrids, one of which, Raubritter, is exceptionally beautiful; unfortunately it is triploid and sterile, so no further generations have been raised. A similar breeders' dead end has been Cerise Bouquet, a hybrid between *Rosa multibracteata* and a Hybrid Tea; Crimson Glory.

Macrophylla Roses–*Search* in Old Roses.

Miniature, Miniflora and Patio Roses–Miniature roses grow to only about 6"-18". The plants, leaves are all miniatures of the larger roses. Miniature roses tend to be quite hardy and can be grown in containers.

Modern miniatures were raised from the dwarf China rose, Roulettii, long grown on cottage windowsills in Switzerland, and rediscovered by Henri Correvon in 1922. Mini-roses are particularly popular at present in America, for growing in small yards or indoors under lights; some of the larger varieties are known as Patio roses. New colours and shapes have been introduced by crossing with Floribundas, and singles by using *Rosa wichuraiana*. Other dwarf Chinas, such as Pompon de Paris were very popular as pot plants in the nineteenth century.

The 1970s also saw the introduction of Miniature Roses, perfect replicas of their big brothers and sisters. These were followed by dwarf Shrub Roses, now generally, (though unofficially) called Patio Roses. Both of these groups are ideal for small gardens and quite lot of people now concentrate on growing them to the exclusion of the taller varieties.

In response to the demands of hybridizers and others, the ARS in May of 1999, approved the official addition of a new classification of roses. This new classification has been given the name Mini-flora.

It is being used to designate the increasingly more popular miniatures with larger flowers and larger leaves, that did not fit into the miniature designation properly, but also did fit into the floribunda classification.

There have already many roses registered in the new classification. Some of these have variously been previously informally designated as patio roses or sweetheart roses by the hybridizers and growers introducing them.

Modern Roses:–Refers to roses introduced since 1867 when the first Hybrid Tea was created. Usually refers to Hybrid Tea, Floribunda, or Grandiflora roses.

Moss Roses–Moss roses originate as mutations or sports on normal roses. The first is recorded in 1720. At present they are known to have appeared three times on Centifolia roses, and less often on Damask roses, in which the moss is stiffer and brownish. A single-flowered Centifolia Moss which appeared in the early nineteenth century enabled hundreds of Moss hybrids to be bred, in addition to the numerous sports that had appeared in the eighteenth century.

Moyessi Roses–*Search* in Old Roses.

Musk Roses–*Search* in Old Roses.

Old Roses (abbrev. OR, OGR, AR)–Sometimes called Old Roses, Old Garden Roses, Old-fashioned Roses or *Antique Roses*, these are the varieties of roses that existed before 1867 when the first Hybrid Tea was introduced. Some of the classes of Old Roses are the Albas, Bourbons, Boursaults, Centifolias, Chinas, Damasks, Gallicas, Hybrid Perpetuals, Mosses, Noisettes, Portlands, and Tea roses. Some of the Ramblers and Rugosas are considered Old Roses.

As a group, Old Roses tend to be once blooming, though some are repeat bloomers. They tend to be more disease resistant and require less maintenance than the Hybrid Teas which accounts for some of their popularity. There are exceptions to this, especially the China and Tea roses. The China and Tea roses are tender and disease prone, but are very important because they provide the repeat blooming genes to many classes of roses (notably Hybrid Teas).

Pimpinellifolia (Spinosissima) Roses–The early Scotch or Burnet roses raised in about 1900 were selections of *Rosa pimpinellifolia*, some possibly crossed with *Rosa pendulina* to introduce red colouRosa Kordes; however, crossed varieties of the Burnet rose with Hybrid Teas, and produced a beautiful range of single-flowered shrubs such as Frufihlingsmorgen (1940) and Fruhlingsgold (1950).

Polyantha Roses–The first dwarf Polyanthas were perpetual-flowering dwarfs of Rosa multiflora crossed with a dwarf China. These were hardy and floriferous, but small-flowered and without scent. In around 1910 the Danish breeder Poulsen began to cross dwarf Polyanthas with Hybrid Teas to try to introduce more hardiness into Hybrid Teas. These were called Poulsen roses, or Hybrid Polyanthas. Few of these survive today, having been overtaken by the

Floribundas which were derived from them.

Noisette Roses–The forerunner of the Noisette rose was a hybrid between Rosa moschata and Parson's Pink China made by Champneys in 1802 in South Carolina and named Champneys' Pink ClusteRosa It was a climber with bunches of semi-double pink flowers but summer flowering only like *Rosa moschata*. Seeds of Champneys' Pink Cluster were raised by Phillipe Noisette in Charleston, and from these he selected Old Blush or *Rosa noisettiana* which was illustrated by Redoute in 1821. This, when crossed with Park's Yellow China, produced the climbing yellow Noisettes such as Desprez a fleur Jaune, yellow Tea roses, and finally, large-flowered climbers such as Marechal Niel.

Patio Roses–Search in Miniature, Miniflora and Patio Roses.

Portland Roses–Portland roses were named after Margaret Cavendish Bentinck, 2nd Duchess of Portland. The original, a hybrid between an Autumn Damask and *Rosa gallica* Officinalis, has been known since 1792. Portland roses were valued for their late flowering as well as their rich red colour, and were soon crossed with Chinas to produce the forerunners of the Hybrid Perpetual.

In the nineteenth century, by crossing roses from the Orient with those from the west, new groups were developed. Among these were the Portland Roses, though few of them now remain.

Poulsen Roses–By the early twentieth century the Polyantha Roses were already popular; but to improve them still further, the Danish Poulsen family crossed some Polyanthas with Teas, thus creating, in 1924, a new group that became known as the Poulsen Roses.

Rambling Roses–Ramblers throw up long non-flowering shoots from near the ground in summer; these shoots produce clusters of flowers along their length the following year and are then finished and in gardens pruned away; Climbers, on the other hand, produce shoots that have a productive life of more than two years and often become very thick and woody. There are three major groups of Ramblers, distinguished by their ancestry:

Rambling Roses: Sempervirens Ramblers–This was the earliest group to be raised, hybrids between the European Mediterranean Rosa sempervirens and other unknown roses. The most famous group such as 'Adelaide d'Oridans' was raised by M. Jacques around 1827.

Rambling Roses: Multiflora Ramblers–These were crosses between the Chinese Rosa multiflora and various Hybrid Perpetuals or Hybrid Teas made around the end of the nineteenth century. Some included *Rosa wichuraiana* in their parentage through 'Crimson Rambler', an old Japanese hybrid. These are mostly quite hardy.

Rambling Roses: Wichuraiana Ramblers–Two very different looking groups belong here; the large flowered, such as Albertine were mostly raised by Barbier at Orleans around 1900-20, and have large flowers similar to loose Hybrid Teas. *Rosa luciae,* a close relative of *Rosa wichuraiana,* was said to be one parent of these, with Teas or Hybrid Teas as the otheRosa

The small-flowered, such as Dorothy Perkins, are said to be a cross between Rosa wichuraiana and a Hybrid Perpetual. Most of these were raised in the USA by Jackson Perkins, or M. H. Walsh. Other Ramblers of generally similar appearance have been produced using different members of the synstylae section, such as Kew Rambler with *Rosa soulieana,* and Baltimore Belle using the Prairie Rose *Rosa setigera.* This last should be exceptionally hardy.

Rubrifolia Roses–*Search* in Old Roses.

Rugosa Roses–These roses are Search indlings or crosses of *Rosa rugosa,* a very hardy species from northern Japan and Siberia. They are all thorny, with leaves with impressed veins, and generally well scented flowers. Several hybrids were raised in the early part of this century of which the Grootendorsts are the most familiar; they were Rosa rugosa crossed with a multiflora hybrid.

Species Roses–Species roses are those found growing in the wild.

Shrub Roses–Shrub roses are close relatives of species roses that have been improved by hybridizers for garden culture.

Sweet Briar Roses–The most familiar of these hybrids of *Rosa rubiginosa* were made by Lord Penzance, using the semi-double form Janet's Pride, crossed with various China hybrids and *Rosa foetida;* these kept the scented foliage of the Sweet Briar which has been lost in later crosses.

Tea Roses–The Tea rose, Rosa x odorata, is a hybrid between *Rosa gigantea* and *Rosa chinensis,* that occurred long ago in China, and many varieties were cultivated there. Two of these were introduced to Europe: the pale pink Humes Blush tea-scented China (1809) and Parks Yellow tea-scented China (1824), but unfortunately neither is now in cultivation in Europe. Their progeny, however, have survived, and a whole class of Tea roses was bred froth them, by crossing with Bourbons and Noisettes, These, when crossed with Hybrid Perpetuals, became Hybrid Teas.

Tea roses are generally either climbers or small, sparse bushes, with a continuous succession of large flowers of great beauty, in shades of pink, buff or light yellow. They have few thorns and are rather tender (American hardiness zone VIT), growing best around the Mediterranean, in California and in Australia.

Chapter 14
Tea, Coffee and Wine Terms

Acerbic–This is nothing but a taste fault in the coffee brew giving an acrid and sour sensation on the tongue. It is the result of long-chained organic compounds due to excessive heat during the holding process after brewing.

Acid–Experts recognize three types of acidity in coffee: 1) natural desirable: acid, 2) natural undesirable: sour, and 3) undesirable: process acidity–generally having with a biting, puckery flavor. Acidity is normally a characteristic of arabica coffees, particularly of high-growth varieties.

Acidity–1. The pleasant tartness, snap, or twist of coffee is known as acidity. It is distinguished from sour, which in coffee terminology means an unpleasant sharpness. Aged coffees, and some old crop, low-grown coffees, have little acidity and taste almost sweet.

2. The crisp or tart taste typically in white wines. This comes from the natural acids in wine: tartaric, malic and citric acid. Wines are described as being high or low in acidity.

Acidy–A primary coffee taste sensation created as acids in the coffee combine with the sugars to increase the overall sweetness of the coffee. Found most often in washed arabica coffees grown at elevations about 4,000 feet, Acidy coffees range from piquant to nippy. A term used to describe a coffee in which this desirable cup characteristic occurs.

Acrid–A secondary coffee taste sensation characterized by a predominantly piercing sour sensation on the posterior sides of the tongue. It is caused by higher-than-normal percentage of sour acids and a high concentration of salts.

Aftertaste–The sensation of brewed coffee vapors, ranging from carbony to chocolaty to spicy to turpeny. Taste is released from the residue remaining in the mouth after swallowing.

Aged–A taste taint that gives coffee beans a less acidy taste and greater body. The result of enzyme activity in the green coffee beans is created by a chemical change during the aging process after harvesting.

Aging–Wine can age in bottles, barrels, vats, or stainless steel tanks. Many wines improve during the aging process, which can take anywhere from five months to five years before the wine is ready to be sold to the public. The best conditions for aging wines are a dark room with high humidity and a stable, cool temperature.

Agony of the leaves–An expression describing the unfurling or relaxation of curled leaves during steeping.

Alcohol–It is an important part of wine. Most wines range from 7 percent to 14 percent alcohol by volume. Alcohol is produced naturally by yeast during the fermentation process. The active yeast converts the natural sugars in the grape juice into alcohol and carbon dioxide, which is released from the fermenting tank by means of a trap.

Alkaline–A supplemental coffee taste sensation characterized by a dry sensation at the back of the tongue.

It is caused by the presence of alkaloid compounds.

Anhui–Anhui is a major black tea producing region of China.

Appellation–The French term, Appellation d'Origine Contrôlée, (AOC), refers to a set of comprehensive regulations that specify the precise geographic area in which a given French wine can be made. AOC regulations also stipulate the types of grapes that can be used, the manner in which the vines must be grown and how the wine can be made. The Italian equivalents of France's AOC laws are known as DOC, Denominazione di Origine Controllata, and a slightly more strict set of regulations known as DOCG, Denominazione di Origine Controllata e Garantita. In the U.S., the regulations governing AVAs (American Viticultural Areas) are far less strict than French or Italian appellation laws. AVAs are designated by the Bureau of Alcohol, Tobacco and Firearms. There are now more than 130 areas that have been designated as AVAs including such well known AVAs as the Napa Valley, Stags Leap District, Russian River Valley, Anderson Valley and so on.

Arabica–"*Coffee arabica*" is the name assigned to the coffee tree by European botanist Linnaeus while categorizing the flora of the Arabian peninsula.

Aroma–1. The characteristic fragrance of brewed tea, imparted by its essential oils.

2. Strictly speaking, aroma can't be separated from acidity and flavor. Acidy coffees smell acidy, and richly flavored coffees smell richly flavored. Nevertheless, certain high, fleeting notes are reflected most clearly in the nose of a coffee, as some tasters say. The best Colombian and Kona coffees are particularly noted for their floral aroma. The sensation of the gases released from brewed coffee, ranging from fruity to herby, as they are inhaled through the nose.

3. The smells from the wine that deal more with what we taste. The human nose can detect over 2,000 odors whereas human tongue can detect only four main groups: sweet, salty, bitter and sour. Common descriptors include: "fruity", "floral" and "earthy".

Aromatic–Designates a coffee that fully manifests the aroma characteristic of its nature and origin.

Assam–A type of tea grown in the state of Assam, India. These are usually black tea and known for their strong, deep red brewed color.

Astringency–1. The drying sensation in the mouth left by teas high in unoxidized polyphenols.

2. This is that mouth-drying effect when we drink red wines. Our mouth puckers and our tongue suddenly feels like velvet. This comes from young or very strong tannins. An astringent wine may be young and in need of aging. It may also simply need to breathe.

Astringent–A secondary coffee taste sensation characterized by a predominantly searing, salty sensation on the anterior sides of the tongue. It is caused by acids increasing the saltiness and typified by an unwashed Indonesian robusta coffee. In regard to coffee, astringency is identified with undesirable acidity.

Autumnal–A term describing tea harvested late in the growing season.

Baggy–An off-taste often observed in cups from weakly roasted coffees that have been stored for a long time in unsuitable conditions.

Baked–A taste and odor taint that gives the coffee brew a flat bouquet and insipid taste. The result of the roasting process proceeding with too little heat over too long a period. Generally unpleasant characteristic of having an over-baked taste in an over-heated coffee. Ranks in the following order of intensity: cooked, baked or burnt.

Bakey–Tea taster expression for over fired teas.

Balanced–Balance appears to mean that no one quality overwhelms all others, but there is enough complexity in the coffee to arouse interest. The Mexican sample should be most balanced, but it has less to balance than the other two coffees. If one tasted the Yemen Mocha against a standard Ethiopian Harrar one should probably sense how the Yemen coffee is similar to the Harrar, but much more balanced. A well-balanced coffee contains all the basic characteristics to the right extent.

Balance–It is the harmony between a wine's acidity, fruits, alcohol and tannin. These should all play an even roll and no one should be more prominent than another otherwise it is known to be flawed. A well-balanced wine is a primary goal of the wine maker.

Barolo–It is both a full-bodied red wine made from Nebiolo grapes and a region in Piedmont, Italy where some of the best examples of this beverage can be found.

Barrel fermentation–White wines fermented in oak barrels as opposed to stainless steel tanks. Stainless steel makes wine more crisp and leaner whereas oak barrels make wines more full-bodied and bold flavors.

Barrel tasting–Barrel tasting is siphoning the wine from the barrel straight to the wine glass. These special events are where barrels are opened and tasted before actual bottling has taken place. Naturally the wines will be young but it allows serious wine enthusiasts the opportunity to investigate limited releases before buying.

Basic tastes–Sweet, sour, salt, and bitter tastes characterized respectively by sucrose, tartaric acid, sodium chloride, and quinine.

Beany–Specific aroma of an insufficiently roasted coffee that has not been able to develop its full aroma.

Bergamot–The essential oil of the bergamot orange, which is mixed with black tea to give Earl Grey tea its characteristic flavor.

Billy–Australian term referring to tin pot with wire handle to suspend over an open fire in which tea is boiled.

Biscuity–A term describing tea that has been well fired, often associated with Assam teas.

Bitter–1. A basic taste characterized by solution of quinine, caffeine, and certain other alkaloids. Perceived primarily at the back of the tongue. Generally normal characteristics of coffees connected with their chemical constitution, influenced by degree of roasting and the method of preparing the brew. Canephora are more bitter than arabica coffees. Up to a certain level, it is a desirable characteristic.

2. Bitterness in wine may be due to a grape variety, an extremely dry climate during the growing season, or poor wine making. Bitterness due to excess tannin.

Black beans–Dead coffee beans that have dropped from the trees before harvesting. Used as the basic unit for counting imperfections in grading coffee on the New York Coffee Exchange. Has a detrimental effect on coffee taste.

Black tea–Prepared from green tea leaves that have been oxidized, or fermented, imparting a characteristic reddish brew. It is the most common type of tea worldwide.

Bland–Lacking coffee flavor and characteristics. A primary coffee taste sensation created as the sugars in the coffee combine with the salts to reduce the overall saltiness of the coffee. It is found mostly in washed arabica coffees grown at elevations below 2,000 feet, such as a Guatemalan. Bland coffees range from soft to neutral.

Blend–1. A combination of different types of teas for flavor consistency from season to season.

2. Blending is a primary task of the wine maker. Wines from different lots or barrels are blended together to produce the final product for bottling. Tradition and regional laws dictate what grape varieties may be blended together to make a certain wine. It is up to the wine maker to select the percentages of each type of grape for the final blend. The particular characteristics of the vintage play a crucial role in this decision. The classic blending example comes from Bordeaux, where by law wine can be made from a blend of the following grapes: Cabernet Sauvignon, Merlot, Malbec, Petit Verdot, and Cabernet Franc.

Blind tasting–It is how those who are seeing-impaired taste wine. It is a means to provide insight into one's true feelings regarding what is being tasted.

Bloom–A term describing the sheen of the tea leaf.

Body–1. A term describing the sense of fullness that the brewed tea imparts.

2. Tea taster's term to describe sheen or lustre present to finished leaf.

3. Body or mouth feel is the sense of heaviness, richness, and thickness at the back of the tongue when one swishes the coffee around one's mouth. In other words, the term is used to describe the mouth-feel of a drink, corresponding to a certain consistency. The coffee is not actually heavy; it just tastes that

way. To follow a wine analogy again, burgundies and certain other red wines are heavier in body than clarets and most white wines. In this case wine and coffee tasters use the same term for a similar phenomenon.

4. The density perception of a wine when it is in one's mouth—either light, medium, or full. A Cabernet Sauvignon is an example of a full-bodied wine; a Sauvignon Blanc is a light–or medium-bodied wine.

Bold–A term describing large leaf cut tea; tea taster's term to denote a full strength brew

Botrytis Cinera–This beneficial mold causes the grapes to shrivel, leaving sugar-laden fruit full of rich, concentrated flavors. Botrytis cinerea doesn't always develop and, when it does, it's sometimes very late, rewarding only those daring vineyard owners who haven't picked their grapes (but who have risked losing the entire crop due to inclement weather).

Bouquet–1. The total aromatic profile created by the sensations of gases and vapours on the olfactory membranes as a result of the volatile organic compounds present in the fragrance, aroma, nose, and after taste of coffee.

2. A French term for the aroma of a wine. Often the first indicator of a wine's quality during wine tasting. Aromas may include fruit, spice, and other smells associated with a particular grape variety, region, or condition of the wine.

Brackish–A taste fault giving the coffee brew a salty and alkaline sensation. The result of salts and alkaline inorganic material left after evaporation of water from the brew due to excessive heat after brewing.

Brassy–A term describing an unpleasant acidic taste, associated with improper withering of the tea leaves.

Bready–Bready taste manifests in coffees that have not been roasted long enough or at a high enough temperature to bring out the flavor oils.

Break–An auction term describing a tea lot for sale, usually at least 18 chests.

Breathing–It is nothing but allowing a wine to mix with the air. Aeration occurs by pouring the wine into a larger container, such as a decanter or large wineglass. Breathing can be beneficial for many red wines and also for some young white wines. Chemically, breathing enables oxygen to mix with the wine, which hastens the aging process. If a wine stands open for more than 12 hours, it will begin to turn to vinegar as the oxygen continues to work. Whether to let a wine breathe before serving depends on the wine. Contrary to popular belief, it is not always beneficial to let older wines breathe prior to drinking, as this can cause them to "turn"–or go bad–before dinner is over.

Brew–Specific taste of a good home brew prepared properly.

Briary–Denotes a wine having an aggressive, prickly taste best described as "peppery". Sometimes the term is combined with the adjective "brawny" to characterize a young red wine with high alcohol and tannin content.

Brick tea–Tea leaves that have been steamed and compressed into bricks; the bricks are then shaved and brewed with butter and salt and served as a soup.

Bright–Denotes a bright red brew or light leaf, as opposed to a dull brown or black color.

Briny–Applies to a coffee that has been over-roasted.

Brisk–A tea high in astringency. It is also a trademarked characteristic of Lipton tea.

Brix–The measurement of soluble solids in grapes at harvest, taken with a refractometer and expressed in degrees. In unfermented grapes, degrees of Brix are approximately the same as percent of sugar. After fermentation, the alcohol concentration is roughly half the sugar concentration of the juice. Thus, grapes harvested at 22.5 degrees Brix will produce a wine with an alcohol content between 12.5 to 13.5 per cent.

Broken–Smaller leaf style usually created during manufacture by passing the leaf through a cutter

Brut–A French term meaning "raw"; describes champagne or other sparkling wines around the world that are very dry, containing no more than 1.5 percent residual sugar.

Buttery–1. A relatively high level of oily material suspended in the coffee beverage. It is due to the presence of substantial amounts of fat in the beans. Most often it is a characteristic of high coffee-to-water ratio brews.

2. A description of a wine, usually a white wine that has taken on a slight buttery flavor. This often happens as a result of the wine being barrel fermented and then left for a period of time in contact with the yeast.

Cabernet Franc–Red wine grape used for blending with Cabernet Sauvignon. It is an earlier-maturing red wine, due to its lower level of tannins.

Cabernet Sauvignon–One of the primary grape varieties and successfully grown in many countries. Cabernet Sauvignon is often called the king of red wines.

Caffeine–A stimulating compound found in tea.

Cambric tea–A weak tea infusion in an excess of milk and sugar.

Canephora–The coffee species second in importance to "*Coffea arabica*," "*Coffea robusta*" is known by botanists as "*Coffea canephora*."

Capsule–The protective metal or plastic sheath over the cork and neck of a wine bottle. The capsule keeps the cork from drying out and admitting air into the bottle.

Caramelized–Corresponds to the taste acquired by roasted beans that have been dipped in sugar, dextrin syrup, or molasses before roasting. It is also perceived in spray-dried instant coffees.

Caramelly–An aromatic sensation created by a moderately volatile set of sugar carbonyl compounds found in coffee's nose that produce sensations reminiscent of either candy or syrup.

Carbony–An aromatic sensation created by a slightly volatile set of heterocyclic compounds found in coffee's after taste that produces either sensations similar to a creosol-like substance or a burnt substance.

Catechins–A class of polyphenol found in high concentrations in green tea, but lower and varied concentrations in other teas derived from the teaplant.

Caustic–A detrimental coffee taste sensation characterized by burning, sour sensation on the posterior sides of the tongue. It is caused by alkaloids increasing the sourness of the acids in combination with a high percentage of salts.

Cellar–It is a storage area for wine, not necessarily underground. A cellar is the best area to keep wines for aging. Ideal conditions are darkness, controlled cool temperature, and high humidity. Bottles should be stored on their sides to keep the corks from drying out.

Ceylon–Teas from Sri Lanka.

Cha–Romanized spelling of the Chinese and Japanese characters for tea.

Chaff (Roasting)–Chaff is paper-like stuff that appears though the roasting process. These little brown flakes are fragments of the innermost skin (the silver skin) of the coffee fruit that still cling to the beans after processing has been completed. Roasting causes these bits of skin to lift off the bean.

Chai–Indian term for tea, often short for masala chai, or spiced tea, which is made from strong black tea combined with milk, sugar, and spices.

Champagne–Only 75 miles northeast of Paris, the region has over 300 villages and produces the best-known sparkling wines in the world. Only wines produced here can legally be called Champagne.

Chaptalization–The process of adding sugar to the fermenting wine to raise the final alcohol level is known as chaptalization. The process is popular in France to raise low alcohol levels by 1 to 2 percent. As the sugar is converted to alcohol, it does not add sweetness to the finished wine.

Chardonnay–This grape is grown in many wine regions around the world. It is responsible for the great white wines from the Burgundy region of France. Chardonnay ranges from medium to full-bodied and is frequently aged in oak barrels to enhance its flavor.

Charmat process–The process of producing sparkling wines in tanks rather than bottles is known as charmat process. It is often used to mass-produce inexpensive sparkling wines.

Chemical–A definite chemical flavor (such as formaldehyde) not to be confused with Rio flavor.

Chest–Classical tea package, usually made of wood and aluminum-lines, used to ship tea from plantation.

Chesty–A term describing tea that has taken on the undesirable smell or taste of the wooden chest in which it was shipped.

Chewy–Related to a wine that has a full, almost thick mouth feels. Zinfandels are often described as chewy.

Chicory–It is a complex bitter-acid and sweetish taste characteristic of the root of the chicory plant.

Chocolaty–An aromatic sensation created by a moderately volatile set of pyrazine compounds found in coffee's aftertaste that produce sensations reminiscent of unsweetened chocolate of vanilla.

Chunmee–It is a grade of Chinese tea with a curled shape.

City or full city roast–"City" is a roast that is slightly darker than the American roasting norm. "Full City" is definitely darker than norm; sometimes patches of oil on surface.

Claret–This term is what the England calls red wines from the Bordeaux in France.

Clean–Without off-flavor.

Cloudy–It indicates the opposite condition of clear or brilliant. It is possibly the result of sediment being stirred up during transportation.

Cloying–Overly sweet, and lacking the correct amount of acidity to give the wine balance.

Color–It may be white, red, and rose each of which having its own spectrum from light to dark. It is the contact between the grape skins and the juice that give a wine its color. During the wine-making process, the longer the juice is in contact with the skins, the more color will be imparted to the wine. A pink or rose wine is made from red grapes but is only allowed brief contact with the skins.

Common–Coffee of ordinary and average quality is said to be as common.

Complex–A descriptive term for a multifaceted, multi-layered wine that continues to reveal different flavors as you drink it. A complex wine, because it is so fascinating, has an almost magical ability to draw the wine drinker in.

Complexity–Complexity describes flavor that shifts among pleasurable possibilities; a harmonious multiplicity of sensation. The Yemen Mocha definitely should be complex; if the Sumatran is a good one it should also be complex; the Mexican is undoubtedly the least complex coffee of the three.

Condrieu–It is an exceptional but hard-to-find white wine from Northern Rhone region in France. The single grape variety used in Condrieu is viognier. Condrieu is a full-body white wine, rich in alcohol, but with a strong flavor and round in the mouth. Condrieu exhales apricot, pear and almond aromas.

Congou–A general term for Chinese black tea, derived from gongfu.

Cooked–A typical taste of an instant coffee treated at too high a temperature.

Coppery–A term describing a bright reddish infusion, associated with black teas of high quality.

Corkage fee–A sometimes small but usually ridiculously high fee a restaurant will charge if one bring one's own bottle of wine to their restaurant. It's been getting more reasonable but one should be aware of it. Typically it's anywhere from $10 and more. It is some what justified in charging a nominal fee but not to that extent. One should call ahead and ask if concerned.

Corked–A wine that is "corked" or smells "corky" is when the seal of the wine bottle's cork is broken. When this happens, certain molds can grow and the aromas from the mold create a 'wet newspaper' smell in the wine. Musty, sweaty shoes, cardboard are frequently used adjectives as well. Also, if air is introduced through the cork, it can oxidize the wine within the bottle also turning the wine. Accidental drinking of this shall not do harm but one must avoid it with sense.

Course–A coffee that is rough on the tongue is said to be as course.

Creamy–Moderately high level of oily material suspended in the coffee beverage. The result of pronounced amounts of fats present in the beans.

Creosol–A supplemental coffee taste sensation characterized by a predominantly scratching sensation at the back of the tongue. It is caused by the high percentage of phenolic compounds created by a dark roast.

Crisp–A wine with a lively acidity level is crisp. A French Chablis, a Sancerre, or a light California Sauvignon Blanc will have the characteristics of a crisp, refreshing wine.

CTC–An acronym for Crush, Tear, and Curl, a machine-based process which macerates the leaves by pressing through counter-rotating rollers to create a stronger, more coloury tea.

Darjeeling–Tea grown in the Darjeeling region of India, near the Himalayas. These (generally black) teas are well known for their crisp astringency.

Dark–Roasting term meaning dark brown beans with a shiny surface; equivalent to espresso or French roast.

Decaffeinated taste–Special process taste often found in decaffeinated coffees. It results due to something lacking or to additional flavors.

Decaffeination process–Coffees are decaffeinated in their green state. Three principal processes are used today: the traditional or European process, the water-only or Swiss-water process, and the CO_2/water or sparkling water process. All are consistently successful in removing all but a trace (2 to 3 per cent) of the resident caffeine.

Decanting–Decanting is the act of pouring the wine from the original bottle into a carafe or other glass container. The reasons for doing so are for two main reasons. One, if you're drinking an unfiltered wine, this is the best way to filter off sediment before serving it. Secondly, decanting also allows the wine to "breathe" through aeration. This helps oxidize the wine releasing it's aromas and flavors.

Delicate–A secondary coffee taste sensation characterized by fragile sweet-subtle sensation just past the tip of the tongue. Caused by the lowest possible combination of sugars and salts that still produce a sweet cast to the taste, a combination easily broken up by other taste sensations. Typified by a washed New Guinea arabica coffee.

Depth–Depth describes the resonance or sensual power behind the sensations that drive the taste of the coffee. It is a tricky and subjective term, but it tries to get at the way certain coffees open up and support their sensations with a sort of ringing, echoing power, whereas others simply present themselves to the palate and then stand pat or even fade.

Detrimental coffee taste sensations–Common to natural coffees that are harsh due to bitter replacing sweet in the taste modulation. The result of sugars being ingested by the shrub as the cherries remain on the branches while drying. Range from medicinal to caustic.

Dhool–A term describing the coppery, fermenting tea leaf.

Dirty–Literally a dirty flavor, not earthy or musty.

Dry–Dryness is a function of the residual sugar in the finished wine–the drier the wine, the less sugar it contains. Most table wines are dry.

Dull–A coffee is dull if it gives an impression of roundness but at the same time lacks character. Dull comes close to the meaning of flat.

Dust–The smallest grade of tea, this is typically associated with lower quality, but is prized for its quick extraction and is commonly used in teabags.

Earl Grey–Black tea that is scented with the essential oil of bergamot, a citrus.

Earthiness–Earthiness is a flavor defect deriving from careless, primitive processing that in some contexts may be seen as virtues. Some Harrar coffees sold in specialty stores may have a hint of wildness or earthiness to them. Roasters from Italy often like to include some earthy-tasting Brazilian coffees in their espresso blends. If a New Orleans blend is at all authentic it also should have some Brazilian wildness in it.

Earthy–1. An odor taint in the coffee beans that produces a dirt-like taste sensation. Results when fats in the coffee beans absorb organic materials from the ground in the drying process during harvesting. Also referred to as dirty and groundy. The undesirable odor and taste of freshly turned soil is found in low-graded batches. Due to poor preparation conditions and botanical origins of the green coffee. Reminiscent of potato flavor also found in instant coffees.

2. A wine tasting term. It means just what it sounds like, a slight taste of soil. The French use a term "gout de terroir" that is often used to mean the same thing; although it can also mean that the wine has typical tastes for the region.

Enology–The science and study of wine and wine making is known as enology. It can also be spelled as **Oenology**.

Estate bottled–Wine that was bottled by the vineyard owner. Many wines are still bottled and produced from grapes that are purchased on the open market, often for the lowest price. This designation assures that the winery had control over the grapes from beginning to end so that they could produce a high quality wine.

Fannings–Small, grainy particles of leaf sifted out of better grade teas.

Fermentation–1. The process of oxidizing green tea leaves to make black and oolong teas.This step involves allowing the natural browning enzymes present in tea leaf to oxidize fresh green tea leaves and to impart the darker brown-red color and characteristic aroma.

2. The transformation of sugar into alcohol and carbon dioxide gas by the action of yeast. The primary chemical process in wine making.

Fermented–A taste fault in the coffee beans producing a highly displeasing sour sensation on the tongue. The result of enzyme activity in the green coffee beans changing the sugars to acids in the drying process during harvesting.

Fibrous–A term describing teas that contain a high percentage of fannings.

Filtering–A wine making technique used to ensure clarity and sterility in wine. Careful use of precise filtering pads and agents allow the winemaker to target specific foreign substances to remove, based on their size. Some winemakers feel that filtering reduces the quality of wine.

Fine cup–Coffee with good, positive characteristics.

Finish–1. If aroma is the overture of the coffee, then finish is the resonant silence at the end of the piece. Finish is a term relatively recently brought over into coffee tasting from wine connoisseurship; it describes the aftertaste that lingers on the palate after the coffee is spit out or swallowed. It is in part a reflection of body; heavier-bodied coffees like the Sumatran will have a much longer finish than lighter-bodied coffees like the Mexican.

2. This is the lingering sensation after we taste a wine. How long the flavors last after it is swallowed is an important factor in great wines. Did it last several seconds or fade quicker than the time it took to swallow? Was it light-bodied (like water) or full-bodied (like the consistency of milk)? In great wines, the finish should last long and and displayflavors well after our sip.

Firing–The process of rapidly heating the leaf, either with hot air or in a wok, to quickly halt fermentation and dry the leaf to its final product.

Flat–1. A term describing teas that lack astringency.

2. An odor taint in the coffee bean or brew meaning that limited range of gases and vapors is present in almost imperceptible strength. Due to aromatic compounds leaving the beans as part of the staling process after roasting or the holding process after brewing.

Flavor defects–Harshness and sourness are two of the most widely used negative epithets. Harshly flavored coffees are unpleasantly bitter, sharp, or irritating. Terms like grassy, hidey, barnyard fermented, musty, and Rioy (medicinal) describe even more dramatically undesirable flavor characteristics. All of these characteristics derive from careless processing. Presumably the coffees you taste will be superior, hence free from such defects.

Flavor–Flavor is the most ambiguous term of all. Acidity has something to do with flavor, and so do body and aroma. Some coffees simply have a fuller, richer flavor than others, whereas other coffees have an acidy tang, for instance, that tends to dominate everything else. One can also speak of a distinctively flavored coffee, a coffee whose flavor characteristics stand out. Of the three coffees the Yemen Mocha is probably the most distinctive, the Mexican the least distinctive, and the Sumatran the richest.

Flight–A selection of wines to taste side-by-side which usually share a common theme for comparison.

Flowery–Used in grading the size of tea, it typically indicates a leaf style with more of the lighter colored tips.

Flush–The freshly-picked tea leaves, typically comprising the bud and first two leaves of the growing tea shoot.

Foreign–A term that generally covers a number of imperfect flavors coming from contamination, for example, rubbery or moldy.

Formosa–Tea produced in Taiwan; primarily Oolong teas.

Fortified–A category of wines to which a neutral wine spirit has been added to create distinctive types of wine and to increase alcohol strength.

Foul–A rank, strong, fermented flavor or any other strong, unpleasant defective flavor, such as hidey or oniony.

Fragrance–The sensation of the gases released from ground coffee as they are inhaled through the nose. Ranges from sweetly floral to sweetly spicy.

French roast–When applied to roasting coffee, means that the bean is roasted high enough to bring the natural oil of the coffee to the surface. Gives a roasted flavor to the cup.

Fresh–A positive characteristic applying to freshly harvested and roasted coffee whose flavor is particularly vivid. An aromatic highlight in the coffee bean and brew that is highly pleasing. The result of extremely volatile organic compounds, particularly those containing sulfur, evoking a strong sensation on the olfactory membranes.

Fruity–An aromatic sensation created by a highly volatile set of aldehydes and esters found in coffee's aroma. Either a sweet sensation reminiscent of citrus fruit or a dry sensation reminiscent of berry fruit.

Full–1. A term describing a strong, vibrant tea infusion without bitterness and posessing good color

2. An intensity description of bouquet indicating gases and vapors are present at a moderately pronounced strength.

Fume Blanc–Simply another name for Sauvignon Blanc. Robert Mondavi used this term as a marketing ploy to play up its smoky flavors.

Gamay–The grape of the Beaujolais region of France. It produces a light styled red wine. While this grape variety is found on the labels of some California wines, it is probable that the grape is not actually Gamay, but a light Pinot Noir clone or even the easily forgotten Valdiguie of southern France.

Genmaicha–Green tea with toasted rice.

Gewurztraminer–Also called Traminer. A perfumed, pungent, spicy, white grape that produces semisweet to dry wines; a specialty of Alsace (France) but also grown in California, Germany, Eastern Europe, Australia, and New Zealand.

Golden–Denoting the orange colored tip present in high quality black tea

Gong fu–A Chinese term meaning performed with care; describes a style of brewing with many repeated short infusions of leaf in a miniature pot.

Good cup quality–Coffee with good, positive all-round characteristics.

Grady–A background flavor of dirtiness but not qualifying as dirty. Mostly used in the United States.

Grainy–A term describing high quality CTC teas.

Grassy–A odor taint giving the coffee beans a distinct herbal character similar to freshly mown alfalfa combined with the astringency of green grass. Created by the prominence of nitrogen compounds in the green beans while the cherries are maturing. Typical taste of unripe beans and of certain freshly harvested coffee batches, corresponding to the beginning of the harvest.

Green–1. Unfermented, dried teas, traditionally found primarily in China and Japan, but becoming increasing popular in the West due to purported health benefits.

2. A taste taint giving the coffee brew an herbal character due to an incomplete development of the sugar carbon compounds in the roasting process. Results from insufficient heat during too short a period. A taste associated with that of a raw fresh vegetable leaf, often found in early new-crop coffees.

Gunpowder–Green tea that has been rolled into pellets, which unfurl in hot water to brew.

Gyokuro–A Japanese term meaning pearl dew, referring to green tea produced from shaded plants.

Hard–1. A term describing pungent tea, often positively associated with Assam (Asom) teas.

2. A secondary coffee taste sensation characterized by a predominantly stinging, sour sensation on the posterior sides of the tongue. Caused by higher-than-normal percentage of sour acids and an insufficient percentage of either sugars or salts. Coffee that strikes the palette by mixed sensation. Bitterness and astringency are not are not enveloped by roundness of body. A

hard coffee is poorly balanced. Indicates the quality of the coffee ranking as a matter of degree from strictly soft, soft, softish, softish/hardish, hardish, hard, Rioy.

Harsh–1. A term describing bitter teas.

2. Acrid. Sensation at the same time bitter and astringent, raspy, and disagreeable. Particularly found in some poor quality robusta coffees. Often due to imperfect beans.

Heavy roast–Coffee beans roasted to a very dark brown, with a shiny surface; equivalent to Italian Roast.

Heavy–1. A term describing a thick,full, deep-colored infusion with little briskness or astringency.

2. A moderately high level of solid material suspended in the coffee beverage. Result of fine particles of bean fiber and insoluble proteins present in pronounced amounts.

Hectare–A metric unit of measure equivalent to 2.471 acres. Wineries in Europe use this term to describe the land area of vineyards. Output of wine is measured in hectoliters per acre. A hectoliter is equal to 100 liters or 26.4 U.S. gallons.

Herbaceous–A green, vegetable smell in wine. For example, Sauvignon Blanc is grassy when subtle, herbaceous when overpowering. It is not considered a positive attribute when it is more than subtle.

Herby–An aromatic sensation created by a highly volatile set of aldehydes and esters found in coffee's aroma. Produces either an sensation reminiscent of an onion or green vegetable.

Hidey–An odor taint that gives the coffee beans a tallowy and leather-like odor. Result of a breakdown of fats in the coffee beans, due to an excessive amount of heat applied in the drying process during harvesting, usually when dried with a mechanical dryer.

Hot–Describing a wine high in alcohol and giving a prickly or burning sensation on the palate. Accepted in fortified wines, but not considered as a particularly desirable attribute in Cabernet Sauvignon or Chardonnay. Positively undesirable in light, fruity wines.

Hydrolyzed–Refers to conventional type of instant coffee having an undesirable acidity due to treatment. Generally associated with over-extraction.

Hyson–Chinese green teas meaning flourishing spring.A brand of tea in common usage during 18th century.

Iced Tea–Tea brewed and served chilled.

Insipid–A taste taint giving the coffee brew a lifeless character, due to a loss of organic material in the coffee bean. Result of oxygen and moisture penetrating the bean fiber after roasting.

Instant taste–Reflects fewer of the organoleptic characteristics that typify home-brewed coffee.

Intensity–A qualitative measure of the number and relative strengths of the gases and vapors present in the bouquet of the coffee.

Italian roast–Term applied to coffee that has been roasted darker than French Roast. Much used by

Italians, as well as in many of the coffee producing countries.

Jasmine–Black tea scented with jasmine flowers typically made with green Pouchong tea as the base.

Keemun–Black tea from central China, typically hand-rolled and fired.

Lapsang souchong–A Chinese black tea which is fired (dried) over a smoky (pine wood) fire to impart its characteristic smoky flavor.

Lees–The sediment that settles at the bottom of tanks after the fermentation process. It is made up of grape seeds, pulp, stems, and skins, and is not transferred when the wine is moved to a different container.

Light–1. A term describing tea liquor lacking body or thickness

2. A moderately low level of solid material suspended in the coffee beverage. Result of fine particles of bean fiber and insoluble proteins present in perceptible amounts.

Malolactic fermentation–The bacterial conversion of the crisper, apple-type malic acid to the softer, milk-type lactic acid in wine. Also called ML or secondary fermentation, this acid conversion yields wines with increased complexity and softer acidity.

Malty–1. A term describing slightly over-fired tea, sometimes desirable.

2. An aromatic sensation created by a moderately volatile set of aldehydes and ketones that produces sensations reminiscent of toasted grains.

Medicinal–A detrimental coffee taste sensation characterized by a penetrating sour sensation on the posterior sides of the tongue. Caused by alkaloids increasing the sourness of the acids without any taste modulation of sweetness.

Medium roast–Coffee beans roasted to the American norm.

Mellow–A primary coffee taste sensation created as salts in the coffee combine with sugars to increase the overall sweetness. Characteristic found most often in washed arabica coffees grown at elevations below 4,000 feet, such Kona coffee from Hawaii. Mellow ranges from mild to delicate.

Merlot–Merlot is traditionally a red wine grape blended into the wines of Bordeaux. In Pomerol and Saint-Emilion, Merlot dominates the wines. In California, Merlot is blended with Cabernet and bottled on its own.

Metallic–A term describing the dry or coppery taste of some teas.

Methode champenoise–French term for the method used to make champagne, which is fermented in the bottle. French champagnes and many other sparkling wines are produced using this traditional French technique. The monk Dom Perignon is credited with inventing this method.

Microclimate–The climate within a small, defined area. Can dramatically affect the character of the wine produced there.

Mild–A secondary coffee taste sensation characterized by a predominantly sweet tingle just past the tip of the tongue. Caused by high concentrations of both sugars and salts. Typified by a washed Sumatran coffee.

Moldy–Coffee may acquire a moldy taste if kept in poor conditions. Moldiness also depends on conditions during the pulping and cleaning of green beans.

Muddy–1. A term describing a dull, brownish or blackish infusion

2. Characterizes a large quantity of particles in suspension in the beverage.

Mulled wine–Red wine that has been mixed with sugar, lemon, and spices, usually including cinnamon, cloves, and nutmeg. Served hot.

Must–The mixture of grape juice or crushed grapes that is fermented into wine.

Musty–An odor taint giving the coffee beans a moldy odor. Result of fats in coffee beans absorbing organic materials from molds on or in contact with the coffee beans during the drying process. Often the result of insufficient or proper drying and aging.

Neutral–A secondary coffee taste characterized by the absence of a predominant taste sensation on any part of the tongue but causing a distinct parching sensation on the sides of the tongue. Caused by a concentration of salts high enough to neutralize both acids and sugars but not enough to provoke a salty sensation. Typified by washed Uganda robusta coffee.

New crop–A taste taint giving the coffee beans a slight herbal character when brewed. Result of an incomplete enzymatic change that ultimately eliminates this taste taint during the aging process.

Nippy–A secondary coffee taste characterized by a predominantly sweet, nipping sensation at the tip of the tongue. Caused by a higher-than-normal percentage of acids being sour.

Nose–1. A synonym for aroma of the tea.

2. The sensation of the vapours released from brewed coffee as they are exhaled while swallowing. Ranges from caramelly to nutty to malty.

Nose–A term used by wine enthusiasts to describe the smell of a wine.

Nutty-An aromatic sensation created by a moderately volatile set of aldehydes and ketones that produce sensations reminiscent of roasted nuts. Characteristic of poor quality beans, that float, remain lighter in color and have a peanut flavor.

Oak–The most popular wood for constructing barrels. Oak imparts flavors and tannin to wines during the barrel aging process.

Oily–A term sometimes used to denote a coffee that has a roasted oily taste due to a high degree of roasting or an oily coffee having a greasy but not rancid taste.

Old–A roasted coffee that has been left for too long changes aroma and acquires a specific and disagreeable flavor. Similar to oldish but with stronger hay-like flavor.

Oldish–A complete lack of freshness. Somewhat flat taste with a slight flavor of hay.

Oniony–Has a flavor of onions.

Oolong–A form of tea characterized by lighter brews and larger leaf styles. This tea is typically understood as a lightly fermented tea, between green and black tea on a continuum.

Orange pekoe–Referring to size of leaf, not quality or flavor, this term indicates a larger-size grade of whole leaf teas.

Ordinary–Below average quality for growth, grade and type. Bland.

Organic–Organic is an important descriptive term in the contemporary coffee world. An organically-grown coffee must be certified by an international agency as having been grown without synthetic chemical fertilizers, pesticides, or herbicides. Somewhat lower yields and the considerable cost of the certification process account for the higher prices demanded for many organic coffees.

Orthodox–A processing method that imitates the larger leaf styles of hand-produced tea.

Oxidized–The smell of a wine that has been overexposed to air. White wines that are oxidized turn dark golden in color, lack freshness, and have an off nose and flavor.

Pan-fired–Tea that is steamed and then agitated in an iron wok over a fire

Papery–Taste that coffee packed in paper bags or prepared in bad quality filter paper may acquire. In instant coffee can be the result of certain processing operations.

Past crop–A taste taint that gives coffee beans a slightly less acidy taste. Result of enzyme changes in the coffee beans during the aging process.

Peasy–A disagreeable taste of very fresh green peas.

Pekoe–A grade of small, whole leaf tea, from the Chinese term baihao, which refers to the white hairs of the new buds on the tea plant.

Periquita–A red wine grape grown in southern Portugal. A medium-bodied, fruity wine that is usually a good value.

Petite Syrah–This red wine grape grown in California by many producers is not related to the French Syrah of the Rhone Valley. Its characteristics are deep color, pepperiness, full body, and good aging potential.

Petite Verdot–Red wine grapes grown in Bordeaux and used for blending with Cabernet Sauvignon. Many producers have been abandoning this grape, since it ripens late and sometimes not at all.

Phylloxera–An insect that attacks the roots of grapevines, killing the vine. Phylloxera is native to the eastern United States. It was transported to Europe on vine roots and plants and caused vine devastation in France around 1863. Usually a vine will die within several years of the attack.

Picey–A character in wine reminiscent of clove, cinnamon, nutmeg, pepper, or other spices. Can be found in both red and white wines.

Pinot Blanc–This white wine grape has in the past been mistaken for

Chardonnay. The two varieties look very much alike. Grown in a variety of regions–Italy, France, California, Germany, Austria–the Pinot Blanc is a light, smooth, easy-quaffing white. Drink young.

Pinot grigio–Grown in Alsace, Germany, Italy, and very successfully in Oregon, this grape variety can produce round, flavorful, dry white wines. Known as Tokay in Alsace, Pinot Grigio in Italy.

Pinot Meunier–Red wine grapes grown in the Champagne region of France. Pinot Meunier is used for blending with Pinot Noir and Chardonnay to add fruit to champagne. Recently the Pinot Meunier varietal is being grown and marketed in Oregon.

Pinot Noir–The red grape that produces all the great red Burgundies. In champagne it is vinified without skin contact to produce a white wine. A difficult grape to cultivate, it can produce some of the most elegant wines in the world. Also good examples in California and Oregon.

Pinotage–A red grape that is a cross between Pinot Noir and Cinsault. Grown in South Africa, it is fermented at higher temperatures and matured in new oak for finesse and elegant berry flavors.

Piquant–A secondary coffee taste sensation characterized by a predominantly sweet, prickling sensation at the tip of the tongue. Caused by a higher-than-normal percentage of acids actually sweet to the taste instead of sour.

Plain–Tea taster's term to denote dull liquor with sour taste.

Plucking–The process of harvesting the tea by cutting the flush from the growing tea plant.

Point–A coffee with good positive characteristics of flavor, body and acidity.

Polyphenols–Astringent compounds found in tea.

Pomace–The mass of skins, seeds, pulp, and stems left in the fermenting vat or cask after wine making. One of the products that goes into the distillation of French marc and Italian grappa.

Poor–Qualifies a coffee of really common flavor.

Port–Port is a sweet dessert wine. The name comes from Oporto on the Douro River in Portugal. Port is an after-dinner drink of very high alcohol content (17 to 20 per cent). They have incredible cellaring/ aging abilities because they are fortified wines meaning brandy and sugars are added during the fermentation process.

Potato–Has an unpleasant taste of raw potato.

Primary coffee taste sensations–Acidy, mellow, winey, bland, sharp and soury.

Process taste–This term reflects a number of defects. Some technological treatment of coffee can develop well-identified off-flavors: cooked, caramelized, cereal, and acrid.

Puerh–A type of tea most notably from the Yunnan province of China. Damp green tea that has been fermented microbiologically to a black leaf.

Pulping–First step after picking in preparing coffee by the wet method. It consists of removing the outer skin. Machines rub away the pulp without crushing the beans.

Pulpy–Strong, pungent, fruit-like flavor from coffee cherry skins.

Pungent–1. A term describing highly astringent tea.

2. Applies essentially to a full-bodied and slightly aggressive coffee.

Pyrolysis–The temperature (around 465F/240C) at which chemical changes in roasting coffee beans cause them to emit their own heat, thus raising the temperature of the roasting chamber.

Quakers–Term applied to unripe, blighted, or underdeveloped coffee beans.

Quakery–A taste taint giving coffee brew a pronounced peanutty flavor. Result of the presence of light colored, underdeveloped, roasted coffee beans. Caused by picking unripe, green, coffee cherries during harvesting.

Racking–The method of pumping out or siphoning of wine from the sediment is known as racking

Rancid–A taste fault giving the coffee brew a highly displeasing taste. The rancid flavor of a roasted coffee is caused by the oxidation of the fats.

Raw–A term describing bitter tea.

Reserve–Our laws are pretty laid back about this. It doesn't mean a whole hell of a lot. I've tasted "reserve" wines that were wines made with more T.L.C from the winemaker and they were very good.

Rich–Intensity description indicating gases and vapors are present at highly pronounced strengths.

Richness–Richness partly refers to body, partly to flavor; at times even to acidity. The term describes an interesting, satisfying fullness.

Riesling–The great white wine grape of Germany. Its good acidity level provides flavorful, crisp wines. Also found in Alsace (France), California, and the Finger Lakes region of New York State. Also known as Johannisberg Riesling.

Rio–With particular reference to Brazils, an iodine-like flavor that can be very pungent.

Rioy–A taste fault giving the coffee beans a highly pronounced medicinal character. Result of continued enzyme activity when coffee beans remain in the fruit and the fruit dries on the shrub. Usually associated with natural processed coffees grown in Brazil. Typified by coffees grown in the Rio district of Brazil.

Roast taste–Terms describing the characteristic collective flavor complex of darker roasts. The acidy notes are gone, replaced by pungent notes combined with a subtle, caramel sweetness. Some people call this often unnamed group of sensations "roast taste" or the "taste of the roast."

Roasty–Relative strength of the natural components of the coffee flavor is modified by the degree of roasting, resulting in high character.

Robusta–High in caffeine and rather bitter. Generally less acid and less aromatic than arabica coffee. Often slightly woody.

Rolling–The process of crushing the leaves to activate certain enzymes and initiate fermentation; also results in the curled appearance of the final tea leaf.

Rose–A light pink wine, dry to sweet, made by removing the skins of red grapes early in the fermentation process or sometimes by mixing red and white wines. Also called "blush" wine.

Rough–A secondary coffee sensation characterized by a predominantly rasping, salty sensation on the palette or tongue, caused by the additive property of salt taste sensations.

Round–A balanced coffee whose basic organoleptic characteristics are just at the right level, with none particularly apparent, giving the impression of roundness.

Rounded–An intensity description indicating a reduced range of gases and vapors is present at a moderately perceptible strength.

Rubbery–A taste fault giving the coffee beans a highly pronounced burnt-rubber character. Result of continued enzyme activity in the coffee bean when it remains in the fruit and the fruit is allowed to dry on the shrub. Usually associated with natural processed robusta coffees grown in Africa.

Salt–A basic taste characterized by solutions of chlorides, bromides, iodides, nitrates, and sulfates of potassium and lithium.

Sancerre–A region in France's Rhone Region known for their Sauvignon Blancs. The typical "smoky" flavor of the Sauvignon grape that is grown in this area, together with part Pinot Noir grapes, characterizes the taste of the Sancerre wines. Sancerre wines are produced on 15 villages. The wines coming from Bue and Chavignol are recognized as the best.

Sauvignon Blanc–A white wine grape planted around the world. In France it is found in Bordeaux, where it is usually blended with Semillon to make a rich styled wine, but with very little varietal characters. The Loire Valley of France is home to the villages of Sancerre and Pouilly-Fume which may produce the truest expression of this grape. California versions are sometimes called Fume Blanc, a relic of a marketing ploy that popularized the grape, but confused the consumer since there is no clear cut difference between wines labeled Sauvignon Blanc and Fume Blanc. New Zealand is another bastion of Sauvignon Blanc. The distinct "freshly cut grass" aroma of Sauvignon Blanc can range from barely noticeable to overwhelming.

Scorched–A odor taint that gives the coffee brew a slight aftertaste of phenolic and pyridine character with an underdevelopment of the caramelization of compounds. Result of applying too much heat and charring the surface of the bean during the roasting process.

Secondary coffee taste sensations–Piquant to nippy, mild to delicate, tangy to tart, soft to neutral, rough to astringent, hard to acrid.

Sediment–This is the stuff that looks like dirt inside the bottle that makes the liquid hazy when held to the light. Its is a combination of minerals, grape skins, dead yeast cells–anything and everything the juice came in contact with along it's fermentation process. Most wineries filter off these particles but some only filter to a certain degree. The sediment can add incredible complexity over years in a cellar left undisturbed. Most people are turned off when they see the sediment–just filtering is necessary before pouring it in a glass. The consumption will not harm.

Self-drinking–A term describing full-bodied tea that does not need to be blended.

Semillion–A white grape varietal most noted for making great wine out of Bordeaux and the Hunter Valley in Australia. Usually offers citrus flavors when made into a dry wine but more often, its turned into sweet, desert-style wines with flavors similar to honey, raisins and tropical fruits. This is because the vines are susceptible to rot (*Botrytis cinera*) which in turn turns each grape into a sugar bomb of fruit.

Sharp–A primary coffee taste sensation created as acids in the coffee combine with salts to increase the overall saltiness. Characteristic found most often in unwashed robusta coffee. Sharp coffee ranges from rough to astringent.

Smoky–A term describing tea fired over an open fire, resulting in exposure to wood smoke.

Smooth–A moderately low level of oily material suspended in the coffee beverage. Result of fats in the beans present in perceptible amounts.

Soft–1. A term describing underfermented tea.

2. A secondary coffee taste sensation characterized by an absence of any predominant taste sensation on any part of the tongue, except for subtle dryness. Caused by a concentration of salts high enough to neutralize the acids but not high enough to neutralize the sugars. Typified by washed arabica coffee from Santos, Brazil.

Soft-sweet–A pleasant clean taste. Denotes a smooth cup free of any foreign flavors, applies particularly to Brazilian coffee.

Sommelier–This person is a restaurant employee who orders and maintains the wines sold in the restaurant and usually has extensive knowledge about wine and food pairings.

Souchong–Large leaf teas harvested from the third and fourth leaf of the tea plant.

Sound cup–A coffee with no particular positive characteristic and without negative characteristics.

Sour–A basic taste characterized by solutions of tartaric acid, citric acid, or malic acid. The unpleasant acidity of a sour coffee cannot be confused with the natural acidity of some coffees in which this quality is prized. Perceived at the tip of the tongue.

Soury–A primary coffee taste sensation created as acids in the coffee combine with salts to increase overall saltiness. Characteristic found most often in unwashed robusta coffees. Soury ranges from hard to acrid.

Spicy–An aromatic sensation created by a slightly volatile set of hydrocarbon compounds in coffee's after taste that produces sensations reminiscent of either wood-spice (cinnamon) or wood-seed (Clove).

Split–A quarter bottle of champagne, containing six ounces. Used frequently on airplanes and trains.

Stale–A taste fault that gives the coffee brew an unpleasant taste. Result of moisture and oxygen penetrating the bean fiber and adversely affecting the organic material that remains in the coffee bean, occurring in the staling process after roasting.

Stalk–A term describing teas that contain pieces of red stalk from poor plucking.

Steen–Another name for Chenin Blanc often used in South Africa. The term "Stein" is also used in South Africa, but refers to a semi sweet style of white wine. However, many "Stein" blends do contain a large percentage of Steen wine.

Stewed–A taste of coffee infusion that has been heated after cooling and lost its initial aroma.

Stinker–A coffee with no particular positive characteristics and without negative characteristics.

Strawy–A taste taint that gives the coffee bean a distinct hay-like character. Result of the loss of organic material from the green coffee beans while in storage, occurring in the aging process after harvesting.

Strong–Coffee giving a pungent impression in the cup, rich in flavor. Developed by roasting or having a consistent mouthfeel.

Structure–How a wine is built. A wine with good structure has the proper proportions of acid, tannin, and fruit that combine to make a well-balanced wine in which all the properties meld together.

Sultana coffee–The dried husks of the coffee cherry.

Supplemental coffee taste sensations–Common to dark roast coffees that are pungent due to bitter replacing a sweet in the taste modulation ranging form creosol to alkaline.

Sweaty–A coffee probably fading to faded, that has been stored for some time in less-than-ideal conditions and results in a distinct sweaty taste.

Sweet–A basic taste characterized by solutions of sugars (sucrose and glucose), alcohols, glycols, and some amino acids, perceived primarily by the tip of the tongue. A trade term to describe coffees free from harshness of Rio flavor or any form of damage.

Sweetly floral–An aromatic sensation created by a highly volatile set of aldehydes and esters that produce sweet fragrance sensations reminiscent of a flower.

Sweetly spicy–An aromatic sensation created by a highly volatile set of aldehydes and esters that produce a spicy fragrance sensations reminiscent of a sweet spice.

Syrah (Shiraz)–One of the great red wine grapes. At home in the Rhone valley of France, it has made its way to Australia where it is known as Shiraz as well as California where it is still known as Syrah.

Tainted–A coffee with a slightly defective flavor.

Tangy–A secondary coffee taste sensation characterized by a predominantly darting, sour sensation along the sides of the tongue. Caused by a high-than normal percentage of sugars, giving the taste almost a fruity sensation. Typified by unwashed India arabica coffees.

Tannin–A misleading term referring to tea polyphenols, which are different than the tannic acid, polyphenols associated with other plants such as grapes.

Tannins–These are the astringent substances found in the seeds, skins, and stems of grapes, as well as in oak barrels, particularly new oak. They are important in the production of good red wines because they provide flavor, structure, and texture and, because of their antioxidant traits which acts like a preservative, they contribute to long and graceful aging. Tannins often give young wines a noticeable astringency that dries the tongue and give a 'puckering' feeling in the mouth. This quality softens as the wine ages, mellows, and develops character. Wines with noticeable tannins are referred to as tannic.

Tarry–1. Tea taster's term for teas that have been fired over smoky flames, imparting a smoky flavor

2. A taste fault giving the coffee brew an unpleasant burnt character. Occurs during the holding process after brewing, a result of condensation and scorching of proteins.

Tart–A secondary coffee taste sensation characterized by a predominantly puckering, sour sensation along the sides of the tongue. Caused by higher-than-normal percentage of sour acids, almost giving the taste a puckering sensation.

Tastevin–Originating in Burgundian cellars for analyzing and tasting wine, a tastevin is a small, shallow silver cup with raised indentations that help reflect the wine's color and exhibit its clarity. It's become customary for a sommelier to wear a tastevin on a chain or ribbon around his or her neck. The Burgundian wine tasting fraternity, Chevaliers du Tastevin, was named after the tasting cup.

Tat–Shelf made of wire mesh or burlap used to spread the leaves out for withering and fermentation.

Terroir–Anything that effects the quality of the grapes grown on the vines that is not able to be controlled by man is considered to be the terroir. Think of as the environment the grapes grow in. Aspects that make up a region terrior are climate, natural irrigation, hillside, soil type, geography, and a myriad of other factors.

Theaflavins–Orange red polyphenols unique to fermented teas such as black tea, and formed from the condensation of two catechins.

Theanine–An amino acid unique to tea.

Theine–A synonym for caffeine.

Thick–A relatively high level of solid material suspended in the coffee beverage. A result of fine particles of bean fiber and insoluble proteins present in substantial amounts. Most often characteristic of espresso-style coffee.

Thief–Glass or metal syringe used for taking wine samples through the bunghole of a barrel.

Thin–1. A relatively low level of solid material suspended in the coffee beverage. A result of fine particles of bean fiber and insoluble proteins present in imperceptible amounts. Lacks body or substance and is insufficiently concentrated and roasted.

2. A wine that is light-bodied, lacks flavor, and is generally light in color.

Ti kuan yin–Meaning iron goddess of mercy, an especially dark and fragrant type of Oolong tea.

Tipped–A taste taint giving the coffee brew a cereal-like taste. Result of heat being applied too quickly in the roasting process, charring the tip of the bean.

Tipping–Charring the end of the coffee bean during the roasting process, by applying an intense heat too quickly.

Tippy–A term describing high quality teas differentiated by the white or golden tips of the leaves.

Tisane–Herbal tea, that is teas produced from the leaves of plants other than the tea plant.

Tuocha–A type of brick tea using pu erh tea pressed into a bowl-shaped brick.

Turpeny–An aromatic sensation created by a slightly volatile set of hydrocarbon compounds and nitrites found in coffee's aftertaste that produces either resinous sensations similar to turpentine or medicinal sensations similar to camphor.

Twist–Before fermentation, the leaves need to be crushed to initiate oxidation. This imparts the curled appearance of the finished leaf.

Twisty–A coffee showing differing negative characteristics in a single cup or from cup to cup. A coffee with unreliable characteristics.

Two and a bud–A term describing the part of the tea plant that is typically harvested, that is, the top two leaves and the bud. See also Flush.

Ullage–This French term–pronounced "ull-idge" in English–refers to the air space in the neck of an unopened bottle of wine. This air bubble will be entirely within the neck of a normal bottle, but older bottles that have lost some of their contents may be described as having a "mid-shoulder fill" or even "low shoulder fill."

Umami–About a century ago, Japanese scientist dubbed the term "umami"–it is the fifth tasting sense beyond sweet, sour, bitter, and salty. It is the "savory" sensation found in foods like: oysters, tomatoes, and mushrooms. This sense puts its emphasis on tannins–wines to appeal to this sense would be mature reds (soft tannins) and certain white wines (those without oak influence).

Unclean–Having off-flavor. Generally depends on the geographic origin of the beans and how they have been treated. A flavor slightly similar to fermenting but without the pungent, rotting taste.

Undefinable flavor–A coffee with an "off" taste that can not be categorized.

Vapid–An odor taint in the coffee brew marked by a loss of organic material that would normally be in a gaseous state in both the aroma and nose of the brew. Occurs during the staling process after the roasting or the holding process after brewing.

Variety–A qualitative description of the gases and vapors present in the fragrance, aroma, nose and aftertaste of coffee's bouquet, which create a complex pattern of sensations of the olfactory membranes.

Verietal–A wine named for the principal grape from which it is made.

Vertical tasting–This involves various bottles of wine but as opposed to 'horizontal tasting' where we are tasting different varietals; in vertical, the varietal is the same but the vintage is different.

Vin de pays–A French term that simply means wine of the region or country. A category of ordinary table wines meant for quick drinking.

Vintage–The term vintage is the period of picking or harvesting grapes each year. It is not when the juice was transported to bottles and distributed. Certain years may be reffered as "vintage years"–these are worthy years to remember because they are considered to have produced high quality grapes ideal for winemaking. It can even be broken down further by getting specific to a certain varietal. For example, for California Cabernets, '99 was considered a vintage year along with '01.

Vintner–A winemaker.

Viognier–The white grape of the northern Rhone Valley of France where it makes the expensive wine known as Condrieu. In the early 1990s, more than thirty top California producers began making viognier to much acclaim. The wine has an opulent, lush body and dramatic honey suckle, white melon and jasmine flavors.

Watery–A relatively low level of oily material suspended in the coffee beverage. Result of slightly perceptible amounts of fats present in the beans.

Weak–Coffee that lacks body but is not flat.

White–A special type of green tea. Distinguished by the presence of the white hairs of the tea flush (baihao) and a lighter green, almost clear, infusion.

Wild–A taste fault in the coffee beans characterized by extreme variation between sample cups. Usually marked by unpleasant sourness. Result of internal chemical changes in the green coffee beans or external contamination.

Winey–1. Mellow quality, characteristic of some Keemun teas which have been given time to age.

2. A primary coffee taste sensation created as the sugars in the coffee combine with the acids to reduce the overall sourness. Characteristic found most often in unwashed arabica coffees grown at elevations above 4,000 feet, such as an unwashed Djimmah from Ethiopia. Winey coffees range from tangy to tart. Special and agreeable flavor acquired by certain mocha-type, freshly milled, or first crop coffees.

Wishy-washy–Negative in all aspects but with no defective flavor.

Withering–The first step in production of most teas. Involves letting the fresh leaves wither for some period of time after plucking to reduce moisture content.

Woody–1. A term describing an unpleasant grass or hay flavour in black tea.

2. A taste fault giving the coffee beans a distinct, unpleasant wood-like character. Result of an almost complete loss of organic material in the green beans during storage. Makes coffee unsuitable for commercial purposes. Reminiscent of the odor of dry wood.

Yeast–The single cell organisms that are responsible for fermentation. This is as true in wine as it is in beer or even bread. In the case of wine, the primary yeast responsible for the first (alcohol) fermentation belong to the class "Sacharomyces." Not all yeast is good yeast, and some can lead to spoilage. Many types of yeast may be found in and around wineries, and due to the need to control the specific yeast in wine, sterility is extremely important in a winery.

Yixing–Pronounced ee-hsing, a region in China known for its purple clay, and the unglazed teapots produced from it.

Yunnan–Tea grown in the Yunnan province, in the southwest of China. These black teas are known for their spicy character. This region also produces Pu-Erh tea.

Zinfandel–A red wine grape found almost exclusively in California. While its origins are undoubtedly European, the exact location and name of its progenitor is one of wine's great mysteries. The Primativo of Italy is often named as the origin of Zin, but some believe that Primativo came from Zin, and not the other way around. Evidence for this is found in DNA testing. Whatever the origin of Zinfandel, it is one of the most planted fine wine grapes in California. These vines are older than most other vines in the area, and for this reason produce some of the most intense fruit. For the uninitiated, Zinfandel means a pink wine. It is important to remember that pink wine can be made from any red wine grape, and that the true red Zinfandel shares little with its pink counterpart. Known for its wide variety of styles, most Zinfandel is not particularly tannic (and does not age well) while having an abundance of fruit, to the point of being "jammy."

Chapter 15
Statistical Terms

Absolute difference–The difference, taken without regard to sign, between the values.

Absolute error–The absolute error of an observation x is the absolute deviation of x from its "true" value.

Absolute frequency–The absolute frequency of a variate, as distinct from the relative frequency, namely the ratio of the frequency to the total frequency of all variate values.

Accuracy–Accuracy in the general statistical sense denotes the closeness of computations or estimates to the exact or true values as contrasted with precision, which refers to reproducibility.

Analysis of variance (ANOVA)–A statistical procedure for analyzing variance from experimental data into different partitioning components.

Annualised data–Annualised data show the value that would be registered if the levels of activity measured for a month or quarter were maintained for a full year. This facilitates comparison of data for different time periods (*e.g.* years and quarters).

Area sampling–A method of sampling when no complete frame of reference is available. The total area under investigation is divided into small sub-areas which are sampled at random or by some restricted random process. Each of the chosen sub-areas is then fully inspected and enumerated, and may form a frame for further sampling if desired.

Autocorrection–Autocorrection is data correction performed by the computer without human intervention. It makes particular use of redundancy. Exclusion (elimination) of incorrect records or substitution of a record or its part by data from other records or the correction base. Auto-correction is generally done according to rules that assure the final (corrected) record fails no edits.

Autocorrelation–The internal correlation between members of series of observations ordered in time or space.

Autoregression–The generation of a series of observations whereby the value of each observation is partly dependent upon the values of those which have immediately preceded it, *i.e.* each observation stands in a regression relationship with one or more of the immediately preceding terms. A scheme of autoregression may be regarded as a stochastic process of a conditional kind.

Bar chart–The graphical representation of frequencies or magnitudes of discrete variates by rectangles drawn with lengths proportional to the frequencies or magnitudes concerned.

Base–A number or magnitude used as a standard of reference. It may occur as a denominator in a ratio or percentage calculation. It may also be the magnitude of a particular time series from which a start is to be made in the calculation of a new relative series–an index number–which will show the observations as they accrue in the future in relation to that of the base period.

Base period–Base period is the period of time for which data used as the base of an index number, or other ratio, have been collected. This period is frequently one of a year but it may be as short as one day or as long as the average of a group of years. Generally used in index number analysis.

Baseline (of a graph)–The horizontal line on a graph corresponding to some convenient basic measurement of the variable represented on the ordinate scale.

Basic statistical data–Basic statistical data are data collected on a regular basis (by survey from respondents, or from administrative sources) by survey statisticians in the national statistical system to be edited, imputed, aggregated and/or used in the compilation and production of official statistics.

BASIC–(Beginner All-Purpose Symbolic Instruction Code) A programming language used in minor data processing tasks. Usually resident on small computers.

Bias–Bias is a term which refers to how far the average statistic lies from the parameter it is estimating, that is, the error which arises when estimating a quantity. Errors from chance will cancel each other out in the long run, those from bias will not.

Chi square test (of goodness of fit)–A statistical test, by means of which one can understand whether the observed frequencies have followed the definite rule of theoretical frequencies or not and is calculated based on the follownig formula:

$$\chi^2 = \sum \frac{(f_0 - f_t)^2}{f_t}$$

where, f_0 is each observed frequency and f_t is its theoretical frequency.

Cluster–A group of contiguous elements of a statistical population, *e.g.* a group of people living in a single house, a consecutive run of observations in an ordered series, or a set of adjacent plots in one part of a field.

Cluster analysis–A general approach to multivariate problems in which the aim is to see whether the individuals fall into groups or clusters.

Cluster Sampling–Cluster sampling is a sampling technique where the entire population is divided into homogenous groups, or clusters, and a random sample of these clusters are selected. All observations in the selected clusters are included in the sample.

Coefficient of variation(CV)–It is a relative measure of dispersion or relative variability equal to 100 times the standard deviation(SD) of a random variable divided by the arithmetic mean.

CV (per cent) = [SD/Mean] × 100

Coherence–Coherence of statistics is their adequacy to be reliably combined in different ways and for various uses.

Completely randomized design–An experimental design having several replicates of each treatment (not necessarily the same number of replicates for all treatment and arranged at random within the whole experimental area.

Confidence interval–A statistical range with a specified probability that a given parameter lies within the range. The range of values is usually based on the results of a sample that estimated the mean and the sampling error or standard error. For example, for the population mean, the confidence intervals can be given by the interval whose lower (*L*) and upper (*U*) limits are as follows:

$L = \mu - Zsx-$

$U = \mu + Zsx-$

where μ is the population mean, Z = standard normal variate, and sx- = the standard error of the mean.

Controlled variable–A factor that is held constant (or nearly constant) or one for whose effect an allowance is made in statistical analysis, so that the effect of some other factor or factors may be isolated, measured and evaluated.

Correlation coefficient–1. A numerical index to measure the closeness of relationship of linear association between series of paired variables under comparison.

2. A correlation coefficient is a measure of the degree to which two variables tend to move together. The coefficient has a value between plus and minus 1, which indicates the strength and direction of association.

Correlation–In its most general sense correlation denoted the inter-dependence between quantitative or qualitative data. In this sense it would include the association of dichotomised attributes and the contingency of multiply-classified attributes. The concept is quite general and may be extended to more than two variates.

Critical difference(CD)–The calculated value employing the standard error of difference of means(SE±) and table 't' values which is just significant at a chosen level of significance (5 per cent or 1 per cent). It is used to compare the divergence of treatment means.

Data analysis–Data analysis is the process of transforming raw data into usable information, often presented in the form of a published analytical article, in order to add value to the statistical output.

Data–Data is a representation of facts, concepts, or instructions in a formalized manner, suitable for communication, interpretation, or processing by humans or by automatic means.

Degrees of freedom–It is a concept which measures how many scores would be free to vary within a set of scores. The number of degrees of freedom is usually one less than the total number of variables in the experiment.

Dependent variable–It is a variable whose magnitude depends on, or is a function of, the value of another variable (or other variables). On graphical representation, the vertical or 'Y' axis is conventionally used for the dependant variable.

Design (Experimental)–It is the arrangement method of experimental plots to minimize the effects of uncontrolled variations in fertility and other natural factors.

D^2 statistic–It is a statistical measure of group distance in a population based on multiple characters. It has its wide use in studying the diversity of genetic materials.

Experimental error–The variations among the experimental units, due to causes other than the assignable causes in the experiment.

First order interaction–It is the interaction between two variables. In factorial experiments, the interaction between two main factors.

Frequency distribution–A graphical, tabular or mathematical representation in the manner in which frequencies of variate are distributed over the range of its possible values.

Hypothesis–It is a postulated solution to a scientific problem which must be tested by experimentation, and discarded if not validated.

Independent Sampling–Independent samples are those samples selected from the same population, or different populations, which have no effect on one another. That is, no correlation exists between the samples.

Mean deviation–A measure of dispersion derived from the average deviation of observations from some central value, such deviations being taken absolutely, *i.e.* without reference to algebraic sign. The central value may be the arithmetic mean or the median. Expressed formally the mean deviation is the first absolute moment.

Mean square deviation–The second moment of a set of observations about some arbitrary origin. If that origin is the mean of the observations the mean square deviation is the equivalent of the variance.

Mean square deviation from any given value–In general, the mean square of a set of values is the arithmetic mean of the squares of their differences from some given value, namely their second moment about that value.

Mean–When unspecified, the mean usually refers to the expectation of a variate, or to the arithmetic mean of a sample used as an estimate of the expectation.

Median–The median is that value of the variate which divides the total frequency into two halves.

Mode–The mode is that value of the variate which is possessed by the greatest number of members of the population.

Non-sampling errors–All types of errors in statistics, other than sampling errors, including data collection errors, reporting errors and data processing mistakes.

Partial correlation coefficient–The linear correlation between an independent variable and a dependent variable from which the net variations associated with other variables have been removed.

Plot–Part or whole of a field on which a specific crop or crop mixture is grown on an experiment is conducted.

Precision–Precision is a measure of how close an estimator is expected to be to the true value of a parameter. Precision is usually expressed in terms of imprecision and related to the standard error of the estimator. Less precision is reflected by a larger standard error.

Quota Sampling–Quota sampling is a method of sampling widely used in opinion polling and market research. Interviewers are each given a quota of subjects of specified type to attempt to recruit for example, an interviewer might be told to go out and select 20 adult men and 20 adult women, 10 teenage girls and 10 teenage boys so that they could interview them about their television viewing.

Sampling Variability–Sampling variability refers to the different values which a given function of the data takes when it is computed for two or more samples drawn from the same population.

Scatter diagram–The diagram of dots obtained by putting the pairs of values (of a bivariate data) in X-Y plane, is called a scatter diagram.

Simple Random Sampling–Simple random sampling is the basic sampling technique where we select a group of subjects (a sample) for study from a larger group (a population). Each individual is chosen entirely by chance and each member of the population has an equal chance of being included in the sample, *i.e.* each member of the population is equally likely to be chosen at any stage in the sampling process.

Spatial Sampling This is an area of survey sampling concerned with sampling in two (or more) dimensions. For example, sampling of fields or other planar areas.

Standard deviation–The most widely used measure of dispersion of a variable/variate introduced by K. Pearson (1893). It is equal to the positive square root of the variance. It is a measure of spread (variation) of a population.

Standard error–1. The positive square root of the variance of the sampling distribution of a statistic.

2. Standard error is the standard deviation of the values of a given function of the data (parameter)over all possible samples of the same size.

3. It is estimated as standard deviation divided by square root of the number of observations in a sample.

Standard error, relative–Search in Co-efficient of variation.

Statistical data collection–Collection of information for the processing and statistical analysis towards drawing meaningful inferences.

Statistical measure–A summary (means, mode, total, index, etc.) of the individual quantitative variable values for the statistical units in a specific group (study domains).

Statistics–1. Numerical data relating to an aggregate of individuals.

2. The science of collecting, analysing and interpreting such data.

Stratified Sampling–There may often be factors which divide the population into sub-populations of homogenous groups/strata and we may expect the measurement of interest to vary among the different sub-populations. This has to be accounted for when we select a sample from the population in order that we obtain a sample that is representative of the population. This is achieved by stratified sampling.

Strip plot design–An experimental design in which two different sets of treatments can be tried in large plots with one set of plots superimposed over the other set at right angles.

Structural data-Data on the basic organizational structure of agricultural holdings that do not change quickly over time, such as farm size or land use.

Subsample–A sample of a sample. It may be selected by the same method as was used in selecting the original sample, but need not be so.

Survey–A survey is an investigation about the characteristics of a given population by means of collecting data from a sample of that population and estimating their characteristics through the systematic use of statistical methodology.

Systematic–This word is frequently used in statistics in contrast to "random" in which a variable or a variate changes following same pattern | rule | law.

Target Population–The target population is the entire group, a researcher is interested in; the group about which the researcher wishes to draw conclusion.

Time series–A time series is a set of ordered observations on a quantitative characteristic of an individual or collective phenomenon taken at different points of time.

Total correlation–Same as *Correlation coefficient*..

Total regression (Simple regression coefficient)–A regression coefficient of zero order, *i.e.* a coefficient which involves only one dependent and one independent variate.

Variable–A variable is a characteristic that may assume more than one of a set of values to which a numerical measure or a category from a classification can be assigned (*e.g.* income, age, weight, etc., and "occupation", "industry", "disease", etc.

Variance analysis–The total variation displayed by a set of observations, as measured by the sums of squares of deviations from the mean, may in certain circumstances be separated into components associated with defined sources of variation used as criteria of classification for the observations. Such an analysis is called an analysis of variance, although in the strict sense it is an analysis of sums of squares. Many standard situations can be reduced to the variance analysis form.

Variance–The variance is the mean square deviation of the variable around the average value. It reflects the dispersion of the values around their mean.

Variate–In contrast to a variable, a variate is a quantity which may take any of the values of a specified set with a specified relative frequency or probability. The variate is therefore often known as a random variable. It is to regarded as defined, not merely by a set of permissible values like an ordinary mathematical variable, but by an associated frequency (probability) function expressing how often those values appear in the situation under discussion.

Weighted average–An average which is obtained by combining different numbers (*e.g.* prices or index numbers) according to the relative importance of each.

Chapter 16
Entomology Terms

Abdomen–The hind part of an insect which is the third division of its body.

Acaricide–It is a chemical which kills mites and ticks.

Acarology–Study of ticks and mites.

Acute dermal toxicity–Poisoning effect caused by a single dose of a chemical(pesticide) absorbed through the skin.

Adfrontal areas–Two narrow oblique plates on head of lepidopterous larvae.

Aedeagus–The copulatory organ of a male insect is known as aedeagus.

Aerosol–When spray droplets (having size 0.1–50 μ.m.) of any chemical is dispersed in the air.

Alate–Possessing wings.

Alimentary canal–A digestive tract present in animals including an insect.

Allomone–A special scent that attracts insects *e.g.* bees and thereby facilitates pollination, sometimes may be defensive *i.e.* toxic and repugnant to potential attackers.

Anal margin–Margin of the wing near the anus of an insect.

Anemotaxis–A tendency of some insects to fly into a current of air.

Ant lion–It is a predatory larva of a winged insect which digs a pit in sandy soil and lies with its large jaws partly exposed as it waits for passing insects particularly the ants. It resembles a dragonfly.

Antenna–A paired segmented appendage of an insect located on the head which functions as a feeler, (Plural–antennae).

Aorta–Aorta is the anterior non-chambered part of the dorsal blood vessel of an insect.

Aphid–A small insect which sucks plant sap causing great damage.

Apical margin–The margin towards outer portion of the wing.

Apodeme–One of the rigid processes forming the insect exoskeleton.

Apodous–Footless.

Apposition eye–An eye of day-flying insects which absorbs oblique rays of light in the pigmented walls of the ommatidia.

Apterous–These are the insects which are wingless.

Apterygota–A subclass of primitively wingless insects.

Arachnida–A class of arthropoda that includes the mites, spiders, tick, scorpions, etc.

Arista–It is a dorsally located large bristle particularly on the antenna of certain Diptera.

Aristate–The antenna which contains arista is known as aristate.

Bait–An edible material that is attractive to the pest, which normally contains a pesticide unless used as a prebait.

Beak–A jointed, protruding mouthpart structure of a sucking insect.

Bipectinate–Bushy or feathery antennae having two rows of comblike processes or teeth.

Bird repellant–These are substances which drive away birds or discourage them from roosting.

Book lung–A respiratory sac containing leaf-like plates usually found in spiders.

Bug–Insects of the order Hemiptera (both winged and wingless), lent the term after used for number of insects.

Bursa copulatrix–It is a tube-like blind structure arising from the anterior end of the midgut in certain insects.

Butterfly–A slender-bodied lepidopteran insect having broad, brightly coloured wings and slender antennae with broad club-shaped ends. Usually the adults are harmless to plants but a few caterpillars can attack plants in the summer causing severe damage.

Campodeiform–Larvae resembling some members of Thysamura of which the body elongate and flattened with legs, antennae and cerci well developed.

Capitate antenna–The antenna having with an apical knob-like enlargement.

Cardo–The basal segment of the maxilla is known as cardo.

Caterpillar–The larva of a butterfly or moth possessing five pairs of abdominal legs located on the 3rd, 4th, 5th, 6th and 10th abdominal segments.

Cephalothorax–A body region consisting of head and thoracic segments, as in spider.

Cerci–A pair of sensory filamentous appendages (resembles to antennae) at the posterior end of the abdomen; (Singular-circus).

Chaetotaxy–It is the arrangement and nomenclature of setae on the exoskeleton of the insect, especially in Diptera.

Chelicerae–Anterior paired appendages in arachnida; modified 2nd antennae of the Crustacea; (Singular-Chelicera).

Chemical control–The method of managing pests using certain chemical pesticides.

Chorion–The shell of an insect egg is known as chorion.

Class–A division of the animal kingdom lower than a phylum and higher than an order, *e.g.* the class insecta.

Clypeus–The sclerites between the frons and the labrum.

Coarcate larva–A larva which resembles to the pupa of diptera; the hibernating stage of a blister beetle larva.

Cochineal insect–It is a coccid of the Mexican Opuntia (prickly pear) from which a red dye is obtained.

Cockchafer–It is a common chafer beetle (also known as May bug) of which both the adults and the larva cause extensive damage to plants. It is hard to control but certain insecticides (*viz.*, HCN) can be effective.

Cocoon–It is a silken covering for the protection of pupa of some kind of insect.

Compound eyes–The lateral convex or hemispherical eyes that are made up of numerous visual units (ommatidia), each of which is represented externally by a facet.

Corium–It is the elongated basal part of the front wing of Hemiptera.

Cornicles–These are a pair of dorsal tubes on the posterior region of the abdomen of aphids which secrete a waxy substance.

Costal margin–It is the anterior margin of an insect wing.

Coxa–Coxa is the basal division of leg which joins it to the body.

Coxal gland–In some insects, an excretory gland opening at the lease of the leg.

Crawler–The active 1st inster of a scale insect is known as crawler.

Cremaster–It is a hook-like structure present on the abdominal end of Lepidopteran pupae.

Crochets–The spines present on the tip of prolegs of caterpillars are known as crochets.

Crop–Crop is a thin-walled dilation of oesophagus in certain insects designed to store foods for sometimes before passing it on.

Cross-resistant–When a pest population which has become resistant to one pesticide also becomes resistant to other chemically related pesticides it is said to be as cross-resistant.

Crustacea–A class of the phylum arthropoda which contains the crabs, lobsters, etc.

Ctenidia–These are the rows of stiff spines on the head and the thorax of fleas.

Cuneus–In Hemipteran insects, it is a triangular piece at apex of corium of the hemelytron, and separated from it by a suture.

Cuticula–It is the outer covering of the body wall of an insect composed of a non-cellular layer of chitin.

Cutworm–A caterpillar which becomes active on night time, feed on plant stems near ground level and hard to control as their presence is usually detected only after the damage has been done.

Cyst–It is actually a resting structure containing the juvenile stage of an organism, such as a cyst of a nematode containing larva.

Day-eye–It is a type of insect eye so named as fitted for use in day time, when light is abundant.

Discal cell–It is the cell located in the basal or central part of an insect wing.

Dorsal ocellus–The simple eye in adult insects, nymphs etc is known as dorsal ocellus.

Dorsal shield (Scutum)–It is the sclerotized plate covering all or most of the dorsal surface in males and the anterior portion in females, nymphs and larvae of hard-backed ticks.

Drones–The male bees are said to be as drones.

Dzjerzon theory–A belief that males of honey-bee are produced from unfertilized eggs.

Ecdysone–Hormone inducing moulting in insects.

Economic pest–It is defined as a pest which causes crop loss to the tune of about 5-10 per cent.

Economic injury level–The density of a pest that causes damage equal to the cost of preventing the damage.

Economic threshold level–The density of a pest at which control measure should be undertaken to prevent an increasing pest population from reaching the economic injury level.

Egg stage–It is denoted by the early life of an insect which is also considered as its first stage.

Elytron–In Coleoptera, Dermaptera and some Homoptera, the horny, veinless front wing is known as elytron; plural-elytra.

Empodium–It is a single, pad-like or bristle-like structure often present between the tarsal claws of insects either between paired pulvilli or alone; plural-empodia.

Endochorion–The inner layer of the chorion of shell of an insect's egg.

Endothorax–The internal framework of the thorax is known as endothorax.

Entomology–The science which deals with the study and control of insects.

Entomophilous–Plant or crop pollinated through insects.

Entomophagous (or insectivorous)–Animals that feed chiefly on insects is known as entomophagous.

Epicuticle–Very thin, sclerotinous, outer layer of the cuticle of an insect or other arthropods.

Epipharynx–A mouth-part structure attached to the inner surface of the upper lip and in a few cases, elongated to form a part of the proboscis.

Eruciform–Cylindrical body with both thoracic legs and prolegs much resembling a caterpillar.

Exarate pupa–It refers to the pupa with appendages free and not glued to the body.

Exopterygota–It refers to the insects passing through a simple and sometimes slight metamorphosis with absence of pupal stages.

Exoskeleton–The solid outer portion of the body of insect is known as exoskeleton.

Facetted eyes–The compound eyes of insects are also known as facetted eyes.

Fenestrae–Fenestrae are the degenerated simple eyes of cockroach represented by ocellar spots.

Flabellum–It is nothing but a lobe at the tip of the proboscis of a bee.

Forewings–The first pair of wings that develop on the middle segment of the thorax are known as forewings.

Frenulum–As observed in many Lepidoptera, it is a bristle or group of bristles arising from the humeral angle of the hind wing and extending underneath the front wing.

Froghopper–It is a type of garden insect whose larva secrets froth.

Frontal gland– A gland in the front of the head of some insects from which poison is discharged *e.g.* termites.

Fulcrum–Chitinous structure in the base of insect rostrum.

Galea–It is the outer lobe of the maxilla, attached to the stipes.

Geniculate antennae–It is actually the elbowed type of antennae.

Glossa–One of the paired lobes of the labium between the paraglossae.

Grease band–A band of sticky or greasy material applied to a stem, as a barrier to insect.

Grub–The legless larva of a beetle or weevil possessing a fleshy body and a tiny head.

Haemolymph–The watery blood of insect is said to be as haemolymph.

Haltere–In the order Diptera, these are small, knobbed organ, one on each side of the metathorax, representing the hind wing; (Plural–halteres).

Hamuli–In the order Hymenoptera there are minute hooks on the anterior margins of the hind wings known as hamuli, with which the front and hind wings are held together.

Haustellum–It is an organ found in Dipteran insects where the distal portion of the proboscis is adapted to suck exposed liquids.

Hemi-elytron–The fore-wings of the bugs in which the basal portion is thickened and the distal portion is membranous; (plural: hemi-elytra).

Hexapoda–The insects are so called because of the presence of six legs in their body.

Hibernation–During certain periods *viz.*, winter months, the insects remain inactive. This is known as hibernation.

Hind-wings–These are the wings developed on the third thoracic segments.

Holometabolous–The complete metamorphosis of insect with egg, larva, pupa and adult stages is known as holometabolous.

Hypopharynx–In certain blood-sucking insects like mosquitoes, an elongated tongue-like lobe is present in the mouth forming a part of proboscis, which is known as hypopharynx.

Imago–It is nothing but the adult insect which is completed its metamorphosis.

Insect–The groups of species belonging to the class insecta and phyllum arthropoda which cause damage to the crop by feeding tissues, cutting or sucking the sap.

Insect parasite–Insects which lay eggs on the body of other insects and feed on them.

Insect predator–Insect which catch other insects and feed on them.

Insect vector–It is an insect that transmits a disease-inducing organism or agent.

Instar–In the entire life cycle of an insect it is the stage between successive molts.

Integument–It is nothing but the outer body covering of an insect.

Jugum–In Lepidopteran insect, jugum is a figer-like process extending from the base of the front wing under the hind wing.

Labellum–In many Dipteran insects, it is the modified tip of the labium; (Plural–labella).

Labium–It is the lower lip of the mouthparts of an insect.

Labrum–It is the upper lip of the mouthparts of an insect.

Lacinia–Lacinia is the inner lobe of the maxilla arising from the stypes.

Lamellate antenna–It is a type of antenna which have plate-like structures at its distal end.

Larva–It is the immature, wingless and more or less worm like stage of an insect's life-cycle after egg and before pupa.

Lateral ocellus–The simple eyes present in the larval stage of an insect are said to be as lateral ocellas.

Leaf beetle–These are small bright coloured leaf-eating beetles belonging to the family chrysomelidae.

Leaf miner–It is a type of small larva which burrow into and eat the internal tissues of the plant leaves causing severe harm.

Maggot–The larva of a fly without leg and without well developed head capsule is known as maggot.

Mandibles–The principle jaws of insects which may be tooth-like (as in chewing insects) and needle-shaped (as in piercing-sucking insects).

Maxillae–These are the second pair of jaws in a mandibulate insect.

Mealy bug–It is a small sucking insect the female of which are often wingless and usually covered with a white mealy layer of wax.

Mentum–It is the distal part of the labium bearing the ligula and palpi.

Mesosternum–It is the ventral sclerite of the mesothorax.

Mesothorax–In winged insects, it is the second (middle) segment of the thorax bearing the middle legs and the first pair of wings.

Metapneustic–Insects which have spiracles only at the posterior end of the abdomen, *e.g.* mosquito larva.

Metasentellum–Sentellum or dorsal cuticular shield covering the hind part of the metathorax.

Metascutum–The largest part of the dorsal cuticular, shield covering the metathorax.

Metathorax–In winged insects, it is the third segment of the thorax bearing the hindlegs and the second pair of wings.

Mite–The sucking insects which belong to the phylum Arthropoda and order Acarina are called as mite.

Moniliform antenna–It is an antenna found in certain insects which is made up of bead-like segments.

Monophagous insects–Insects which restrict their feeding to a single species of a food plant are known as monophagous insects, such as, potato tuber moth.

Myriapoda–The class of insects comprising of millipedes and centipedes are known as myriapoda.

Notum–It is the dorsal surface of a body segment in insects.

Nymph–It is a stage of the life cycle of an insect.

Ocellus–simple eyes in adult insects occurring single or in groups.

Ommatidium–A single visual unit of the compound eye.

Ootheca–In certain orthoptera the outer covering of an egg mass is known as ootheca.

Osmeterium–In the anterior body regions of certain butterfly caterpillars a sac-like structure present provided with scent glands. It is known as osmeterium.(Pl–Osmeteria).

Oviposition–Process of egg laying.

Ovipositor–Organ at hind end of abdomen in female insects, through which eggs are laid.

Palpi–Palpi are the segmented feeler-like processes in the mouth parts of an insect bearing tactile and olfactory sense organs.

Paraglossae–The two outer lobes of labium of an insect are known as paraglossae.

Pectinate antennae–It is a special type of antennae which possess one row of projection like a teeth of a comb.

Pedipalpi–The second paired appendage of an arachnid is said to be a pedipalpi.

Phobotaxis–Reflex action lay which an insect turns to avoid an adverse stimulus and by repeated trial and error eventually moves away from it.

Plumose antennae–It is actually a feather-like antennae.

Prepupa–It a quiescent stage between the larva and the pupa.

Proboscis–It is the prolonged mouthparts of an insect specially adapted for piercing, sucking or other type of feeding.

Proleg–In case of caterpillars and sawfly, fleshy and short, unjointed abdominal or false legs are known as prolegs.

Pronotum–The dorsal surface of the prothorax is said to be as a pronotum.

Propodium–In Hymenoptera, the first abdominal segment joint to the thrax is called propodeum.

Propolis–A resinous material of vegetable origin used by bees in addition to wax for the building and repair of their nests.

Prothorax–Among the three thoracic segments of an insect the first one is known as prothorax.

Pupa–It is an inactive, non-feeding, transition stage between the larva and the adult stage in insects with complex metamorphosis. (Pl–Pupae).

Rheotaxis–Response whereby an aquatic insect moves in a definite direction in relation to the flow of water.

Rodenticide–A chemical that can kill or destroy rodents.

Saltatorial legs–Longer, thicker, powerful and specially adapted hindlegs used for jumping *e.g.* grasshoppers, flea beetles etc.

Sawfly–A fly-like insect whose grubs are very much harmful for the garden crops and particularly the females are having with a pair of saw-like swrrated blades in the egg-laying organ.

Semilooper–Certain Lepidopteran larva that crawls by forming small loops of the body.

Setaceous antenna–A type of antenna that tapers towards tip.

Sex pheromone–The chemical with special odour secreted by one sex of an insect to attract the opposite sex for mating purpose.

Stypes–A part of the maxilla.

Stomodeum–In Arthropoda, the forgut is known as stomodeum.

Stylet–A tubular hollow sucking organ of aphids used to suck the sugary content of the phloem vessels.

Tarsas–In insect,the distal division of leg attached to the tibia(Plural-tarsi).

Tibia–In insect, the fourth division of leg between the femur and the tarsus.

Trochanter–In insect, the second division of leg between the coxa and the femur.

Trophallaxis–Mutual exchange of food between insects of a colony *e.g.* of bees, wasps and particularly ants.

Ventriculus–It is the stomach of an insect.

Chapter 17
Instruments, Implements and Tools Used in Horticulture

Auxanometer–The instruments for measuring growth of plants.

Backpack sprayer–It is used for spraying chemicals in vegetables and seed spice crops.

Ball barrow–It is a kind of wheel barrow where a large plastic ball replaces the wheel.

Bill hook–A hatchet with a curved point, used for trimming of hedges, etc.

Budding and Grafting knife–This is used for budding and grafting. It has two soft blades in which one is with ivory edge used for lifting the bark in budding operation

Carton stapler–A heavy-duty, hand-held or mechanical stapler for fastening cartons together prior to filling or for closing the lids after filling.

Chisel–It is a garden implement having one or more soil-penetrating points, with sufficient weight to force the points into the soil for loosening the sub-soil.

Combine planter–An implement which drills out fertilizers, plants, underground parts and covers in the one operation.

Crowbar–It is a steel rod (1.2-1.5 m in length and 2.5 cm in diameter) having one end pointed and flat, generally used for digging soil particularly before laying out a lawn.

Digging fork–This has prongs of 20 cm long fitted to a wooden handle. This is used for uprooting of plants, rooted cuttings, harvesting of tubers etc., without damaging the root system or tubers.

Disc harrow–It is a special harrow with 8–10 high carbon steel discs, fixed in two gangs with separate levers for regulating the angle of discs. The instrument is fitted for work with 1-2 pairs of bullocks by placing on the axle with spacers and bearings.

Disc separator–It is an instrument consisting of a series of cast iron discs which revolve together on a horizontal shaft inside a cylindrical body.

Double knife–Two budding knives or razor blades separated by a wood spacer and securely fastened as a unit for use in patch budding.

Draw hoe–A swan-necked hoe with a broad flat blade set at right angle to the handle. It is used to break up rough ground and to earth up plants by drawing the tool towards the user, rather than pushing it backward and forward like other hoes.

Drum mixer–A simple mixer consisting of a drum, through which a pipe is run at an angle, mounted on two saw horses. It is generally used for dry treatment of seeds.

Duster–Duster is a device through which specific pesticide is applied as dry powder after mixing in a diluent such as talc.

Edging tools A variety of tools used to trim and maintain the edges of lawns, e.g, straight-bladed types of spade, right-angled long-handled shears, electric trimmers, etc.

Face shield–It is a transparent piece of protective device used by a pesticide applicator to protect his face from exposure to pesticides.

Flower gatherer–It is a special secateur designed to hold a flower after cutting it.

Flower pot–Flower pot is a small bucket-shaped container for growing plants with drainage holes in the base.

Fork hoe–A garden hoe often used to break down rough soil and loose the surface of hard soil, and with three prongs set at right angles to the handle.

Forklift tractor–It is a tractor equipped with a mast and forks at both ends for loading, unloading and transporting pallet boxes at roadside near an orchard or grove or at a packing house.

Fruit gatherer–1. It is a mechanical device consisting of a pair of clippers on a long pole with a bag of similar receptacle fixed just below the clipper head to catch the fruit from tall trees without any damage.

2. This is provided with a long handle and a net like structure for holding the harvested fruits. The handle is very light in weight usually with hollow bamboo and the net is made up of ordinary cotton thread or nylon rope.

Fumigun–It is a device similar to a needle for hand-injection of different fumigant chemicals into the soil.

Garden fork–It is used to loosen the soil while harvesting bulb crops like onion and garlic and also in weeding.

Garden rake–This is used for levelling lands and collecting weeds in nursery. The rake consists of a number of nail like projections from a crow bar provided with long handle.

Garden shears–This is used to prune hedges and edges.

Gelmeter–It is a pippetelike glass apparatus for measuring the amount of pectin present in fruit extract.

Grafting knife–A single-bladed knife basically used for grafting of plants with wooden or plastic handle of which the blade is longer and heavier than that of a budding knife.

Gravity seperator–Seed processing machine used to seperate the seeds in different lots on the basis of their differential specific gravity.

Grecian saw–A narrow-bladed curved saw used for cutting out small tree branches.

Gyro sifter–A machine in which grains of smaller size (collected at the centre) are seperated from the larger ones (moving towards the periphery) by passing them over gyrating screens.

Hand fork–It is device similar to a fork but with shorter handle used for working close to the soil.

Hand hoe–It is used for manual weeding.

Hand sprayer–It is nothing but a small portable sprayer used for spraying pesticides, etc.

Hedge cutter–It is a device often mounted in the tractors for trimming hedges.

Hydrometer–It is an instrument for measuring specific gravity used for determining TSS (total soluble solids, expressed in ^{0}Brix) in fruit juices.

Hydro-pneumatic sprayer–Sprayer in which the liquid carried in pressure tank and spraying pressure developed by engine-powered air compressor, either skid type or mounted on wheels; not commonly used.

Hygrothermograph–It is a device for recording temperature and relative humidity simultaneously and continuously.

Hypsometer–It is an instrument which is used to determine the height of standing tree from some distance.

Infiltrometer–A device with which the rate and amount of water percolating the soil is determined by measuring the difference between the amount of water applied and that which runs off.

Iron pan–It is used for lifting plants in nursery. It is used to measure the soil, FYM and sand in pot mixture preparation.

Jelmeter–It is an instrument used for testing pectin during the preparation of Jam, Jelly etc which directly indicates the amount of sugar needed per litre of extract.

Knapsack sprayer–It is a spraying tool often carried on the back of an operator for the application of liquid pesticides in which the attached hose has a nozzle at the tip that can be aimed at the portion to be sprayed.

Knife–It is a device for cutting, layering, budding or grafting plants in a garden. It may be of several types, *viz.* pruning knife, dissecting knife, budding knife, grafting knife, etc.

Lance–It is a long tube fixed to a sprayer which allows the operator to spray at a distance from him or her.

Lawn rake–It is a fan-shaped and wire-, plastic, or bamboo-made garden tool used for removing cut grass, leaves and other debris.

Light meter–It is an instrument for measuring visible light.

Long-arm pruner–It is a pruning instrument operated by a remotely controlled lever mechanism with a long handle.

Lopper–It is a device for trimming trees and shrubs with a long handle.

Lysimeter–It is a device which, under controlled condition, is used to measure percolation and leaching losses from a column of soil.

Mist blower–It is a sprying device in which hydraulic atomization of the liquid at the nozzle is aided by an air blast past the source of spray.

Mister–For increasing humidity of the atmosphere, a fine spray of liquid (water) is provided to the indoor plants by an instrument known as mister.

Moisture meter–It is an instrument for measuring the amount of moisture in the soil, consisting of a probe attached to a calibrated dial.

Mower–It is a machine which is basically used to trim the grasses of a lawn.

Needl epoint holder–It is a stem-anchoring device consisting of many sharp-pointed spike or nails basically used in flower arrangement.

Nozzles–It is device which is used to control drop size, rate, uniformity, thoroughness and safety of a pesticide application.

Onion hoe–It is actually a draw hoe used for weeding in restricted areas and having with a short handle.

Pickaxe–It is a simple tool having double-ended blade with a stout tapering point on one end and a relatively narrow, slightly curved, flat-pointed blade on the other fixed with an wooden handle; normally used for digging or loosening soil.

Plant shredder–It is a device used for tearing limbs, leaves and other plant debris into small pieces and conveying chopped up material into a dump truck or into a pile.

Planting tool–The tools required for planting such as a spade(for planting larger trees, shrubs and herbs), a trowel (for planting smaller plants) etc.

Pneumatic seperator–A device by which seed seperation is made by air on the basis of difference in terminal velocity.

Pot clip–Pot clips are nothing but bended wires which are used to provide support to hold the plant in a pot until it becomes established.

Potometer–Potometer is a device used to measure transpiration (or more appropriately water absorption) of plants.

Power tiller–A walking type tractor usually fitted with two wheels only. The direction of travel and its control for field operation is performed by the operator, walking behind the tractor.

Pressurizing fan–It is a fan placed at the end of the clear plastic greenhouse, which forces heated air, exterior cold air or interior warm air through the tube depending on whether the system is being used for heating, cooling or air circulation, respectively.

Prong cultivator–It is along handled, hoe-like instrument with three or more arched prongs instead of a blade.

Pruning knife–It is a small, blade-shaped knife sometimes with fold-away blades.

Pruning saw–This is used for pruning dead branches of the trees. Small saw with slightly curved edges are useful for removing thick branches and water shoots which could not be sheared off with secateur.

Pruning shears–It is a big scissor-like implement having 20-25 cm blades with a pair of wooden handle used for trimming hedges, shaping plants and even to trim grasses when lawn mower is not available.

Pulper–It is a machine used for chopping up whole fruits and screening out the larger partcles including seeds.

Pyranometer–Pyranometer is used to measure broadband solar irradiance on a planar surface and is a sensor that is designed to measure the solar radiation flux density from the field.

Pyrheliometer–It is a type of instrument which measures intensity of direct

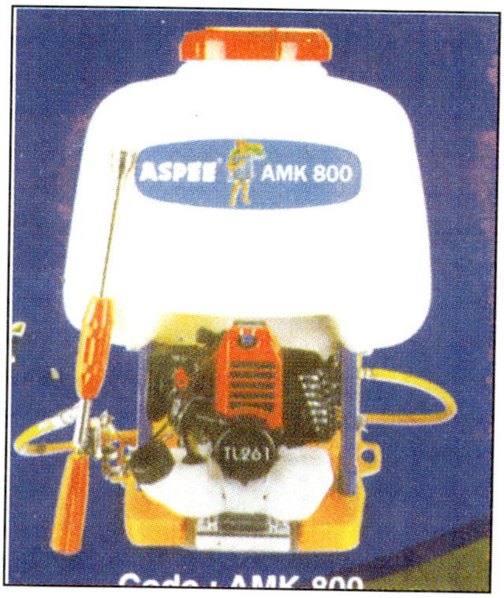

Manual operated

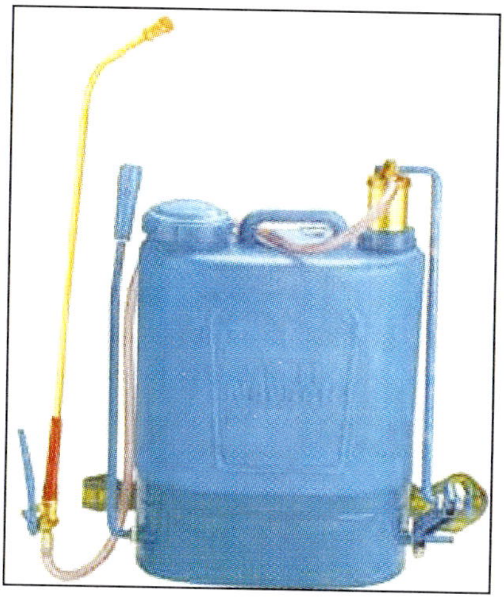

Power operated

Knapsack Sprayer

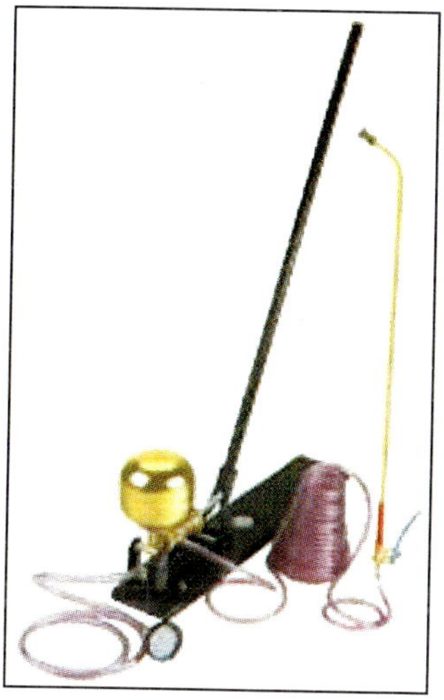

Rocker Sprayer

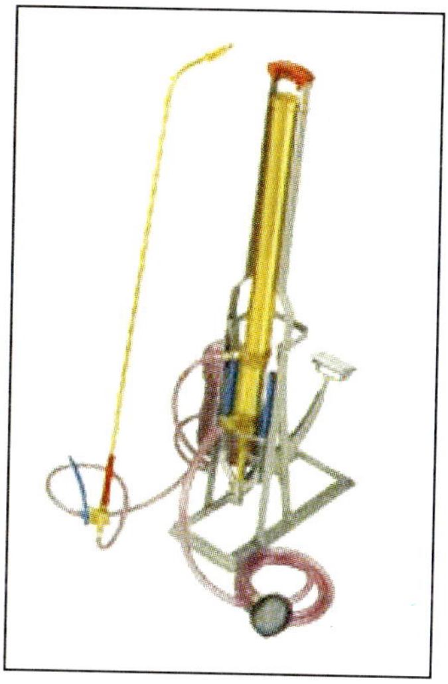

Foot Sprayer

Spin Disc Sprayer
(Battery Operated)

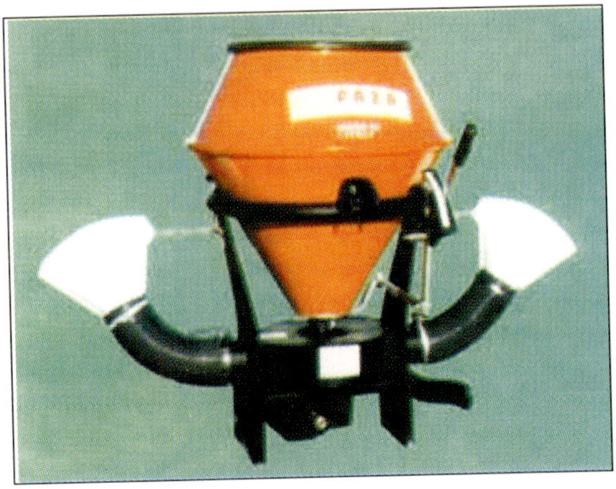

Powered Knapsack Air Assisted Duster and Tractor Operated Spin Disc Duster

Kitchen Sprayer Continuous Pumping

Sprayer with Blower (Electrically Operated)

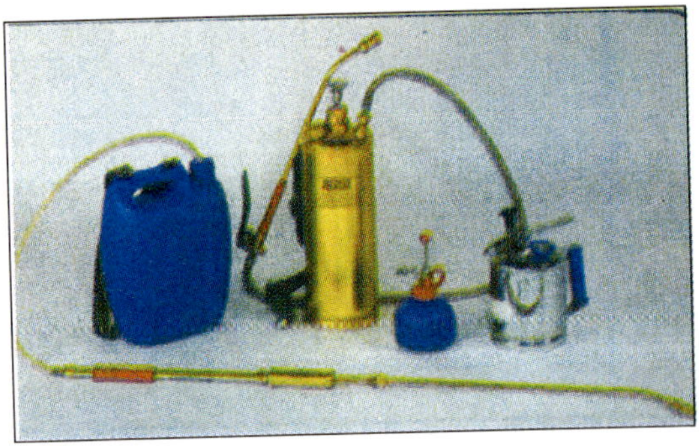

Kitchen Sprayer Continuous Type and Pressure Sprayer

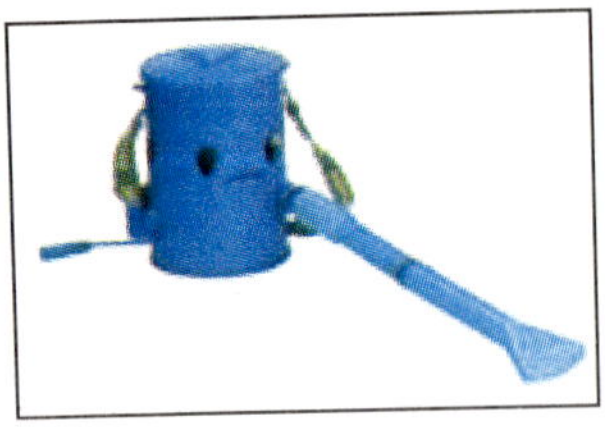

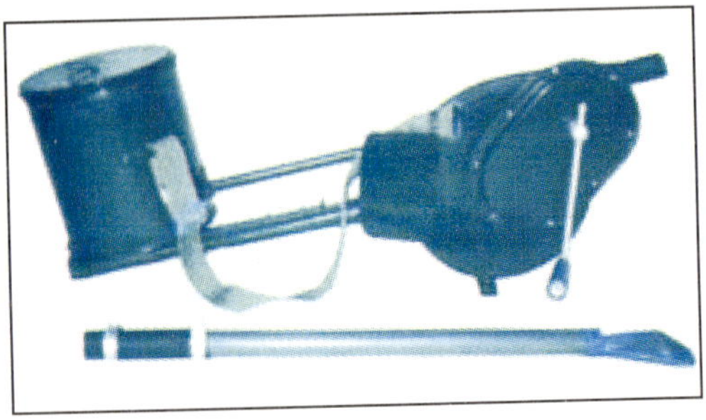

Hand operated dusters

solar radiation (sunlight) perpendicular to the receiving surface.

Rain gauge–It is a type of instrument used to measure the depth of water caused by precipitation distributed over a horizontal impervious surface without any evaporation.

Rake–It is a garden tool containing many teeth or prongs set in a crosspiece attached to a handle, mainly used for cultivating the soil.

Reel mower–It is a grass cutting machine having a series of blades that clip the grass by a scissor action.

Refractometer–This instrument measures the refractive index of liquid, *viz.* fruit juices, with calibration in °Brix (to determine sucrose or total soluble solids).

Rocker sprayer–It is used for spraying chemicals in tall trees.

Roller–It is a basic garden tool used for the preparation and leveling of lawns, seed beds,etc;consisting of a cylinder mounted on a frame with a pushing handle.

Rotary cultivator–A tilling or soil-breaking device with rapidly revolving blades or claws.

Scythe–It is a long flat metal of 5 cm wide with 45–50 cm length fitted with a wooden handle. At the end of the metal, it is slightly curved with sharp edges. This is mainly used for cutting grasses manually.

Secateur–This is used for cutting small shoots to regulate shoot growth in fruit trees, shrubs and vines. It is mainly used for preparation of cuttings for propagation purpose.

Shovel–This is a curved steel plate attached to a wooden handle and used for transferring soil and manure etc.

Sieve–Sieve is a device with a meshed (perforated) bottom that is designed to allow the passage of liquids or fine solids while retaining coarser material.

Spade–An iron square plate fitted to a wooden handle of 30–45 cm length at 45° angle. This is used for formation and rectification of irrigation channels, formation of ridges and furrows, earthing up operation and sometimes even in weeding operations.

Siphon–A hose pipe used to transfer liquid from higher to lower elevation is known as siphon.

Solubridge–A special device which measures the electrical conductivity of the soil solution.

Tractor–Self-propelled power unit, having wheels or tracks for operating agricultural implements and machines including trailors.

Tree pruner–It is provided with a long handle and is used for pruning stray branches which cannot be reached easily.

Weeder–A handtool used for weeding in a garden.

Wheelbarrow–A type of cart used for transferring soil, plant, container, etc, from one place to other of a garden in quantities one person can move.It consists of a metal or wood body with a wheel at one end and handles and legs on the other.

Chapter 18
Bonsai and Flower Arrangement Terms

Bonsai–(In Japanese, *Bon* means shallow container and *sai* means planting); It is the art of growing miniature plants in shallow pot or tray thereby recreating natural landscape and plants for greater aesthetic gratification but requiring minimum of space.

Cascade style bonsai–A style of bonsai in which the trunk of the plant is trained to hang over the edge of the container as if it hangs over the edge of a mountain.

Clustered bonsai–A type of bonsai where several branches are allowed to grow from ground level in a clustered manner.

Dai-kengai bonsai–*Search in* **Kengai bonsai.**

Flower arrangement–It is the arrangement of flowers with leaves or other materials, viz, bouquet, garland, wreath, etc.

Fuki-nagashi bonsai–A bonsai which represents a few storm-bent trees in a cluster growing a top a mountain range along the sea coast, imitating the effect of wind and weather.

Gnarled style bonsai–A type of bonsai where the plants' trunks are twisted near the ground once or twice to form a knob or a loop.

Gohon-yose bonsai–This bonsai allows to grow two or more similar or dissimilar plants in a pot and accordingly named as nihon-yose (when two trunks planted together) or gohon-yose(when five trunks planted together) and so on.

Hankan bonsai–A bonsai with gnarled and twisted trunk suggesting a tree growing on a storm-swept crag.

Ikada-buki–This is raft like bonsai in which a young, well-branched tree is planted horizontally in a container.

Ikebana–(Japanese *ikeru* means 'to keep alive' and *bena* means 'flowers'). It is a specialized and unique Japanese art of flower arrangement regarded by and known to the entire world for its aesthetic appeal, simplicity and economic values.

Jardinière–These are the container or stand basically used indoors for displaying the growing ornamental plants.

Jiyu-bana–It is a Japanese method of flower arrangement in both Moribana and Nezeira styles where one can use combinations of wood, metal or any other material alongwith flower.

Kabumono bonsai–When a single bonsai is present in a pot, it is said to be a kabumono bonsai.

Katate-mochi bonsai–It is a bonsai of height 18-30 cm, grown in small containers; can be carried in one hand, usually placed in small tables or window sills and can be moved from one place to the other.

Kengai bonsai–It is a style of bonsai resembling a waterfall and in which a tree is rooted in a high crevice looking down a deep gorge.

Mame bonsai–It is a type of bonsai which is of very low height (below 18 cm). It is also known as **miniature bonsai** or **baby bonsai**.

Mitsugsoku–It is a style of flower arrangement, which uses a combination of candles(with lampstand) and incense-sticks(or burner) apart from the usual vase with flowers-deepadhar, dhupadhar and pushpadhar-the three essential items of worships according to the custom followed in India.

Moribana–It is a type of flower arrangement where shallow containers supported with pin holders are used to keep the flowers and branches.

Morimono–It is a Japanese style of flower arrangement in combination with fruit and / or vegetables.

Moyo-gi bonsai–It is a free upright style of bonsai in which the irregular upright growth is traditionally arranged in three tiers to represent heaven, earth and man.

Nageire–It is a style of Ikebana (meaning *thrown in*) where tall upright vases are used for making the arrangement of flowers with sufficient stem length.

Ne-agari bonsai–It is characterized by a style of bonsai whose roots raise high above the soil.

Ne-tsuranari–It is a type of bonsai representing several trunks in the pot with connected roots.

Nihon-yose bonsai–Search in *Yose-ue* bonsai.

Oblique style bonsai–It represents a bonsai style in which the plant is grown in an oblique position as if swept by wind and consequently having more branches on one side.

Saba-miki bonsai–It represents an ancient tree with decayed trunk and torn branches. (*Syn.*–Split trunk bonsai).

Semi-cascade bonsai–It is a slanting trunk bonsai imitating a tree growing at right angles to a mountain slope.

Shakan bonsai–Search in Semi-cascade bonsai.

Shito bonsai–A bonsai made of tiny plants in 2.5 cm pots that can actually be held on the tip of a finger (Syn. Finger-tip bonsai).

Sho-kengai bonsai–Search in Kengai bonsai.

Tokonoma–A bonsai with a height range of 30–75 cm.

Wind swept bonsai–Search in Fuki nagashi bonsai.

Yose-ue bonsai–When two or more plants are grown in a pot they are known as yose-ue bonsai *i.e.* trunks planted together.

Nihon yose–Two trunks planted together.

Gohon yose–Five trunks planted together.

Zen-ei-bana–A Japanese style flower arrangement where a beautiful sculpture is created using wood, stone,rocks or metal depicting a sea side or any other natural scenary.

Zen-ei-ka–A Japanese style flower arrangement where straight plant materials of uneven height are used in combination with any other inert material that a person can conceive of.

Chapter 19
Natural and Synthetic Chemicals Used in Horticulture

Aflatoxin–The carcinogenic toxin produced by a few strains of the fungus *Aspergillus flavus*.

Agrochemicals–These are basically the biologically active chemicals widely used in horticulture, *viz.*, growth substances, herbicides, pesticides, fungicides, weedicides etc.

Alar–Succinic acid 2, 2-dimethyl hydrazide, a plant growth regulator, is commercially known as alar which can improve the colour and texture of apple, grape and tomato and prevent permanent dropping.

Allethrin–It is a relatively harmless (particularly for the warm-blooded animals) plant derived (pyrethrin) synthetic insecticide.

Allicin–The active principle for the characteristic pungency in garlic having bactericidal properties. It is an enzymatic cleavage product from its precursor allin, a naturally occurring amino acid.

Alliin–A colourless, odourless, water-soluble amino acid present in the fresh garlic bulb which when crushed breaks down to allicin in presence of an enzyme allinase.

Allomones–Substances excreted by the plant which include recovery and pollination attractants.

Amine–These are nitrogen containing organic compounds which is derived from ammonia.

Anatoxin–A toxin modified by heat or chemical treatment so that its toxic properties are lost but its antigenic properties retained.

Anionic surfactant–Most wetting agents, the surface additives to the pesticides, having a negative surface charges and showing better performances in cold, soft water.

Auxin–Plant growth regulators which stimulates cell division and cell enlargement in plants.

Bacteriocins–Bacterial substances produced by certain strains of bacteria which are active against some other strains of the same or closely related species.

Betalains–Garden beet containing chemicals with red and yellow pigments which can be hydrolyzed into sugar and a coloured portion.

Biological insecticide–Biological agents(viz, *Bacillus thuringiensis*) that can kill insects like a chemical insecticide and then rapidly dissipates in the environment are known as biological insecticide.

Biotin–It is a component of vitamin B-complex, also known as vitamin H having the properties of curing baldness.

Botanical pesticide–A few chemicals viz, nicotine, pyrethrum, strychnine and rotenone which are produced by and extracted from plants functioning as a pesticide.

Bromelin–A protein digestive enzyme present in the mature pineapple fruit.

Capsaicin–Colourless, crystalline, pungent principle($C_{18}H_{27}O_3N$) of chilli which is secreted from the outer wall of the fruit. Actually it is the condensation product of 3-hydroxy 4-methoxy benzylamine and decylenic acid.

Capsanthin–It is a carotenoid pigment which is responsible for the orange-red colouration of ripe chilli.

Carcinogen–An agent or substance that incites or produces cancerous growth is said to be a carcinogen.

Centrifuge–An apparatus used to seperate or remove particulate matter suspended in a liquid by centrifugal force produced by spinning the sample.

Chemosterilant–The chemical used to make an insect sterile.

Chlorophyll–Green, light trapping pigment, essential as an electron donor in photosynthesis.

Chloropicrin–It is a chemical used for pasteurizing root media.

Choline–It is an amine derived from lecithin decomposition process. Both choline and its more active derivative acetylcholine are found in nervous system.

Colchicine–A chemical having the ability to cause chromosome doubling and thereby to produce polyploids.

Conicine–It is an alkaloid usually found in *Amorphophallus* sp which in combination with calcium oxalate causes acridness.

Contact pesticide–Chemicals designed to kill insect-pest when it touches them.

Cross–contamination–The accidental or chance mixing of one pesticide with another usually occurring in a pesticide container or in a poorly cleaned sprayer.

Cucurbitacin–The chemical responsible for the bitterness of certain cucurbits which is basically a tetracyclic triterpene with extensive oxidation level.

Cumarine–It is an organic compound known as tonka bean camphor with characteristic vanilla-like odour.

Curcumin–It is a chemical which is said to be the colouring principle ($C_{12}H_{20}O_6$) of turmeric.

Cutin–Cutin is a group of waxy substances which are deposited both within the epidermal cell walls and on the outer surface of the walls and minimizes water loss.

Dextrin–Being related to starch, this chemical is a more or less a complex carbohydrate.

Diosgenin–It is an alkaloid having medicinal value which can be extracted from the *Dioscorea* sp.

Enterotoxin–A toxin produced by certain strains of *Staphylococcus aureus* during their growth which causes the commonly occurring food poisoning.

Enzymes–An organic material that possesses the ability to promote certain chemical reactions or some physiological life processes without itself being appreciably used up in the process.

Ethylene (CH_2=CH_2)–It is a gaseous plant hormones present in plant tissues which accelerates the ripening of fruit, dropping of leaves and fruit and aging of the plant.

Exoenzyme–An enzyme produced by a plant and exuded into the surrounding area where it breaks down potential food material.

Exotoxin–It is a toxin excreted by anaerobic, spore-forming bacteria (*viz., Clostridium botulinum*) outside its cell into surrounding item (food) and consists of glycoprotein.

Eugenol–An aromatic liquid compound occurring in a number of different essential oils including oils of cloves, pimento, star anise and citronella.

Flavonoids–Chemically, these are 15-C compounds (C_6-C_3-C_6), often synthesized in the chloroplasts, consisting of two aromatic rings linked by an aliphatic three carbon chain; namely flavones, flavonols and anthocyanins.

Formalin–An aqueous solution of formaldehyde is commercially known as formalin.

Formazan–It is a chemical produced during the tetrazolium test which imparts red colour to the living seeds. It is actually an insoluble red compound produced from 2, 3, 5–triphenyl tetrazolium chloride.

Forskohlin–It is a diterpenoid present in the tuber of *Coleus forskohlii* (Chinese potato), useful in certain pharmaceutical preparations prescribed for glaucoma and low blood pressure.

Fructose–Also named as fruit sugar or levulose, it is actually a monosaccharide sugar (carbohydrate) present in fruit nectar and honey.

Fungicide–It is a chemical *e.g.*, Bordeaux mixture, captan, zineb etc, to kill, control, inhibit or manage fungi and often used as disinfectant, eradicant or as protectant.

Gibberelin–It is a specific plant growth regulator responsible for promoting cell elongation and shoot growth. It may be of various types namely GA_1, GA_2, etc.

Glucose ($C_6H_{12}O_6$)–One of the simplest sugars, the ingredient of carbohydrate and the principal product of photosynthesis by the plants.

Glucosans–Starch yielding only glucose on hydrolysis. It is commonly found in cereals, potatoes, legumes etc.

Guanine–It is one of the purines, the basic nitrogenous compound related to uric acid and represented by G.

Inulin–It is a type of carbohydrate very closely related to starch.

Isoprene–One of the substance from which terseness and ultimately essential oils are formed.

Kinin–It is a kind of hormone which is responsible for cell division and other important processes.

Lecithin–It is a fat-soluble compound containing nitrogen and phosphorus.

Lectin–It is a plant protein that binds to specific carbohydrates.

Lycopene–It is a chemical containing red pigment found in tomato.

Melanin–It is an organic (brown) pigment which is responsible for the black-amber dark-brown deposition in the cuticle of insects.

Moringin–It is an alkaloid present in *sojne* plant (*Moringa* sp) which constricts blood vessels, raises blood pressures and accelerates heart beat.

Morphine–It is an alkaloid present in raw opium(latex) which is collected from incisions made in the unriped capsules(fruit) of poppy(*Papaver somniferum*). It is a powerful analgesic, narcotic and having medicinal values.

Mycotoxins–These are the toxins produced by moulds. Of the various toxins, aflatoxin is important. This is a highly toxic, carcinogenic, widespread contaminant of stored foods like spices, groundnut etc and produced by the fungus *Aspergillus flavus*.

Naringin–It is a glucoside located in the tissue containing orange juice sacs, inner portions and in the seeds; the presence of which causes the juice to taste bitter.

Neurotoxin–It is a type of endotoxin produced by *Clostridium botulinum* which affects the peripheral nervous system by adversely disturbing the production of acetyl cholin necessary for conduction of nerve impulses.

Nicotine–A tobacco-derived chemical often used as insecticides specially to control the aphids.

Nucleic acid–It is an acidic chemical containing pentose, phosphorus and bases (pyrimidine and purine) for determining the genetic properties of living organisms chief types are RNA and DNA.

Opium–It is a brownish,solid chemical containing several alkaloids(viz, morphine, thebaine, narceine, papavarine, etc) and obtained by incising the fully swollen capsule (fruit) of opium poppy (*Papaver somniferum*).

Papain–Milk of raw papaya fruit in dry condition is called papain. It is the most important bi-product of papaya, collected by incising green, unripe but well-developped fruits and extensively used as a useful digestant.

Paraffin–A colourless, odourless and tasteless fatty substance obtained through dry distillation from wood, coal, petroleum, used as illuminant, lubricant etc. in sealing tops of jars bearing preserved fruit or similar products.

Pectin–An important chemical useful to prepare jam, jelly, etc, most abundantly found in the middle lamella between cell walls of certain fruits. It is actually a methylated polymer of galacturonic acid.

Penicillin–Small group of closely related chemicals obtained from the clear liquid of a culture of the mold *Penicillium notatum*. It is useful for the treatment of many infections.

Pentosan–It is a kind of polysaccharides which, on hydrolysis, produces five-carbon sugars.

Pheromone–It is a special type of secretion by an insect which attracts other insects of the same species, particularly of different sexes, and influences their behaviour.

Phycobilins–In algae, a special type of red and blue biliproteins, *viz.*, phycoerythrins and phycocyanins are found which are known as phycobilins.

Picoline–It is a basic derivative of pyridine.

Piperine–In Piperaceae family (*e.g.*, Black pepper, Betelvine), an alkaloid is present known as piperine, which is responsible for the unique biting taste of the respective crops.

Quercetin–It is a chemical pigment responsible for imparting colour in the outer skin on onion bulb.

Rhizocaline–A complex of orthodihydroxy phenol, auxin and an enzyme usually present in the cotyledons, leaves and buds which stimulate the rooting in cutting.

Saponin–The bitter principles of certain fruits and vegetables.

Sinigrin–Certain vegetables (especially those belongs to the genus *Brassica*) contains a glucoside which produces allyl-iso-thiocyanate having typical pungent flavour on hydrolysis under the influence of myrosinase enzyme.

Solanine–A poisonous chemical (alkaloid) formed in certain solanaceous crops like potato (which causes greening also)when the tuber is exposed to light.

Solasodine–A glycoalkaloid present in brinjal which is responsible for bitter taste and off flavour.

Sucrose–A non-reducing sugar *viz.*, cane sugar, beet sugar etc, with the empirical formula $C_{12}H_{22}O_{11}$.

Tannin–A chemical having an astringent or bitter taste.

Theobromine–An important alkaloid having medicinal values and present in certain plantation crops viz, cocoa,tea,etc.

Tomatine–A steroidal glycoalkaloid found in tomato, brinjal,etc having antibiotic activity against a number of organisms.

Tryptophan–A type of amino acid a precursor of IAA and serotonin.

Vanillin–The fragrant aromatic constituent of vanilla used to flavour ice-cream etc.

Xanthophyll–A yellow-colored pigment of chloroplast (hydroxyl carotene derivative) which occurs in green leaves together with chlorophyll and carotene.

Chapter 20
Some Medicinal/Medical Terms

Adaptogen–Substance which assists in the adaptation to physiologic or psychological stress, usually through its effect of the hypothalamic-pituitary-adrenal axis. *Example*: Ginseng, Eleuthero

Alterative–Chemical having the effect of restoration of health, *Example*: Goldenseal and Red clover.

Analgesic–Chemical which reduces pain, *Example*: Calendula, Willow bark, St. John's wort and Peppermint.

Anesthetic–1. Substance which is capable of producing anesthesia (loss of sensation and usually of consciousness without loss of vital functions artificially produced by the administration of one or more agents that block the passage of pain impulses along nerve pathways to the brain).

2. Agent that causes unconsciousness or a loss of general sensation.

Anodyne–Chemical that soothes pain. *Example*: Passionflower, Jamaican dogwood, California poppy.

Antiallergic–A substance that reduces tending to relieve or control allergic symptoms.

Antibacterial–An agent that destroys bacteria or suppresses their growth or reproduction.

Anticatarrhal–Substance that helps the body eliminate excess mucus. *Example*: Echinacea.

Antidepressant–A substance which can reduce symptoms of depression. *Example*: St. John's wort, Lemon balm, Lavender.

Antidote–The chemical which actually counteracts a poison is known to be an antidote. *Example*: Milk thistle.

Antifungal–A substance destroying fungi or inhibiting their growths effective against fungal infections.

Anti-inflammatory–1. Agent that counteracts or suppresses the inflammatory process.

2. Chemical that reduces the inflammatory response to illness or injury. *Example*:St. John's Wort, Arnica.

Antimicrobial–An agent which assists in the destruction of pathogenic bacteria, fungi, viruses and parasites or increases the body's resistance to disease caused by these organisms. *Example*: Garlic, Goldenseal, Bloodroot, Wild indigo, Echinacea and Thuja.

Antioxidant-A substance which inhibit harmful oxidation processes in the body. *Example*: Blueberry.

Antiperspirant–An agent having an inhibitory action upon the secretion of sweat.

Antipruritic–A substance having the capability of reducing itching, *Example*: Calendula.

Antirheumatic–A substance that can reduce and prevent joint symptoms. *Example*: Meadowsweet, Black cohosh and Sarsaparilla.

Antiseptic–1. Chemical that kills pathogenic micro-organisms, especially on the skin, Example: Rosemary, Birch leaf.

2. Chemical opposing sepsis, putrefaction, or decay; preventing or arresting the growth.

Antispasmodic–A substance which reduces spasm of smooth muscle tissue. *Example*: Fennel, Licorice, Lobelia, Cramp bark and Wild yam.

Antitussive–An agent that soothes coughs. *Example*: Cherry bark.

Aperient–Chemical that has mild laxative properties. *Example*: Licorice.

Aperitive–An agent that stimulates the appetite.

Astringent–1. A substance able to contract cell walls and stop unwanted discharge.

2. Binds to mucous membranes and reduces irritation and secretions. *Example*: Oak bark, Blackberry root, Nettle, Witch hazel.

Bitters–Bitter tasting herbs used to stimulate gastric secretions, including bile flow, and gastrointestinal motility. *Example*: Motherwort, Gentian.

Carminative–1. An agent that expels or causes the expulsion of gas from the alimentary canal so as to relieve colic or griping.

2. Agent reducing gastrointestinal irritation, inflammation, and gas. *Example*: Chamomile, Rosemary, Dill.

Cathartic–A substance that promotes strong bowel evacuation. *Example*: Senna, Bloodroot.

Cholagogue–1. An agent that promotes the flow of bile into the intestine, especially as a result of contraction of the gallbladder.

2. A substance that carries off bile. *Example*: Blue flag, Yellow dock, Spindle tree.

Choleretic–1. An agent that promotes bile secretion by the liver.

2. Substance that helps increase bile secretion by the liver. *Example*: Burdock.

Cicatrizant–An agent that induces scar formation at the site of a healing wound.

Demulcent–A chemical which can reduce tissue inflammation through high mucilage content. *Example*: Mallow, Licorice, Slippery elm.

Depurative–A substance which is used as a cleanser of impurities. *Example*: Wild indigo.

Diaphoretic–1. An agent capable of inducing sweating.

2. An agent that stimulates sweating. *Example*: Elderberry, Pleurisy root.

Diuretic–1. An agent that increases the excretion of urine.

2. Agent which increases urinary output. *Example*: Dandelion leaves, Parsley, Goldenrod, Bearberry.

Dyspepsia–Impairment of the function of digestion.

Dysphagia–Difficulty in swallowing.

Dyspnea–Difficulty in breathing.

Emetic–Chemical which causes vomiting (often at higher doses). *Example*: Bloodroot, Lobelia, Wild indigo.

Emmenagogue–Substance that stimulates menstruation and normal uterine function. *Example*: Black cohosh, Motherwort.

Emollient–An agent that soothes the skin. *Example*: Calendula, Comfrey.

Escharotic–Agent that irritates the skin and produces a scab. *Example*: Bloodroot.

Estrogenic–Chemical which has effects similar to estrogen, a female hormone. Also called a phytoestrogen. *Example*: Coffee, Soy, Wild indigo.

Expectorant–1. An agent that promotes the discharge or expulsion of mucus from the respiratory tract.

2. Agent that stimulates coughing and clearing of lung secretions. *Example*: Horehound, Lobelia, Anise.

Galactagogue–An agent that promotes the secretion and flow of milk. *Example*: Chaste tree, Goat's rue, Anise.

Glucosuria–The presence of glucose in the urine.

Hypnotic–Chemical which promotes sleep. *Example*: Hops, Passionflower, Wild lettuce and Valerian.

Hypotensive–A substance that reduces blood pressure. *Example*: Hawthorn, Motherwort.

Immunomodulator–Agent that strengthens the immune system. *Example*: Echinacea and Astragalus.

Laxative–Substance that causes evacuation of the bowels. *Example*: Cascara.

Nervine relaxant–An agent which soothes the nervous system. *Example*: Catnip, Lemon balm, Passionflower.

Nervine stimulant–Substance which stimulates the nervous system. *Example*: Mint, Lobelia, Ginger.

Nervine tonic–Substance which strengthens and supports the nervous system. *Example*: Oats, St. John's wort, Hyssop, Motherwort, Skullcap.

Parturient–An agent that aids in childbirth. *Example*: Partridgeberry, Raspberry leaf.

Pectoral–A medicinal substance for treating diseases of the respiratory tract.

Purgative–A very strong laxative. *Example*: Senna.

Rubefacient–1. A substance for external application that produces redness of the skin.

2. An agent which causes redness of the skin due to increased blood flow. *Example*: Horseradish, Cayenne

Sedative–1. An agent tending to calm, moderate, or tranquillise nervousness or excitement.

2. Agent that has a calming effect on mind and body, Example:Hops, Kava-Kava.

Sialogogue–A substance which stimulates saliva secretion. *Example*: Cayenne.

Spasmolytic–An agent tending or having the power to relieve spasms or convulsions.

Stomachic-An agent that strengthens the stomach and excites its action.

Tonic–A substance which may be regarded as nutritive and supportive to whole body or part of body. *Example*: Nettle, Cleavers.

Heart tonic, Cardiotonic: Hawthorn, Motherwort

Liver tonic, Hepatic: Milk thistle

Uterine tonic: Chaste tree

Vasodilator–Chemicals that relaxes arteries in the peripheral circulation. *Example*: Ginkgo.

Vermifuge–An agent that clears body of worms, Example: Wormwood

Vulnerary–A substance which is having the capability to promote wound healing. *Example*: Chickweed, Comfrey, Plantain.

Chapter 21
Some Special Terminologies Collected from Different Sources

A. BIODIVERSITY

[1]**Biodiversity**–The variability among living organisms from all sources including, *inter alia*, terrestrial, marine and other aquatic ecosystems and the ecological complexes of which they are part; this includes diversity within species, between species and of ecosystems (*Source*: The Convention on Biological Diversity. Article 2, UNEP 1992).

[2]**Biological diversity or biodiversity**–It is the total variability within all the living organisms and the ecological complexes they inhabit. Biodiversity has three levels–ecosystem, species and genetic diversity–reflected in the number of different species, the different combination of species and the different combinations of genes within each species. 5-10 million species are estimated to exist, and many of them have millions of genetically distinct individuals. About 1.7 million species have been described worldwide with flowering plants constituting 14 per cent of this total.

Diversity within species is the main concern of genetic resources programmes and the key to the survival of species in nature in the long-term. Drastic reductions in biodiversity are not a new phenomenon in nature. What is new is the scale on which the erosion of biodiversity is occuring. Reduction of diversity will have serious and irreversible consequences for human welfare. (*Source*: IPGRI. 1993. Diversity of Development. The Strategy of the International Plant Genetic Resources Institute).

[3]**Biological diversity = biodiversity**–The variability among living organisms from all sources and the ecological complexes of which they are part; this includes: Ecosystem diversity: the variety and frequency of different ecosystems. Species diversity: the frequency and diversity of different species. Genetic diversity: the frequency and diversity of different genes and/or genomes. It includes the variation within a population and between populations. (*Source*: Schmidt, L. 1997. Tree Improvement Glossary. Technical Note 46. Danida Forest Seed Centre).

[4]**Biological diversity**–The full range of genetic diversity (species, subspecies, and distinct biological populations of plants and animals) as well as the full variety of ecosystems in which the plants and animals occur. (*Source*: Koski, V.; Skroppa, T.; Paule, L.; Turok, J. 1997. Technical guidelines for genetic conservation of Norway spruce (*Picea abies* (L.) Karst.). EUFORGEN. IPGRI).

[5]**Biological diversity**–The variety and complexity of species that are present and that interact in an ecosystem, plus the relative abundance of each. (*Source*: Temperate Forest Foundation. Glossary. USA. http://forestinfo.org/Glossary).

[6]**Biological diversity or "biodiversity"**–It is the totality of genes, species, and ecosystems in a region. The wealth of life on earth today is the product of hundreds of millions of years of evolutionary history. Over the course of time, human cultures have emerged and adapted to the local environment, discovering, using,

and altering local biotic resources. Many areas that now seem „natural" bear the marks of millennia of human habitation, crop cultivation, and resource harvesting. The domestication and breeding of local varieties of crops and livestock have further shaped biodiversity.

Biodiversity can be divided into hierarchical categories–genes, species, ecosystems, and culture–that describe quite different aspects of living systems and that scientists measure in different ways. (*Source*: World Resources Institute (WRI) "What is Biological Biodiversity" http://www.wri.org/biodiv/biodiv.html).

[7]**Biodiversity (biological diversity)**–The variety, distribution, and abundance of different plants, animals and micro-organisms, the ecological functions and processes they perform, and the genetic diversity they contain at local, regional or landscape levels of analysis. Biodiversity has five principal components: (1) genetic diversity (the genetic complement of all living things); (2) taxonomic diversity (the variety of organisms); (3) ecosystem diversity (the three dimensional structures on the earth's surface, including the organisms themselves); (4) functions or ecological services (what organisms and ecosystems do for each other, their immediate surroundings, and for the ecosphere as a whole (*i.e.* processes and connectedness through time and space); and (5) the abiotic matrix within which the above exists (the unity of the soil, water, air, and organisms, with each interdependent on the continued existence of the other). Source: Dunster, J.&K. 1996. Dictionary of Natural Resource Management. CAB International.

[8]**Biodiversity**–The wide diversity and interrelatedness of earth organisms based on genetic and environmental factors. (*Source*: Hagedorn, S. A. An Agricultural and Environmental Biotechnology Annotated Dictionary. http://gophisb. biochem.vt. edu/resources/glossary.html).

[9]**Biodiversity**–1. The variety and abundance of life forms, processes, functions, and structures of plants, animals, and other living organisms, including the relative complexity of species, communities, gene pools, and ecosystems at spatial scales that range from local through regional to global–*syn*: biological diversity 2. An index of richness in a community, ecosystem, or landscape and the relative abundance of these species–note 1. There are commonly five levels of biodiversity: (a) genetic diversity, referring to the genetic variation within a species; (b) species diversity, referring to the variety of species in an area; (c) community or ecosystem diversity, referring to the variety of communities or ecosystems in an area; (d) landscape diversity, referring to the variety of ecosystems across a landscape; and (e) regional diversity, referring to the variety of species, communities, ecosystems, or landscapes within a specific geographic region–note 2. Each level of biodiversity has three components: (a) compositional diversity or the number of parts or elements within a system, indicated

by such measures as the number of species, genes, communities, or ecosystems; (b) structural diversity or the variety of patterns or organizations within a system, such as habitat structure, population structure, or species morphology; and (c) functional diversity or the number of ecological processes within a system, such as disturbance regimes, roles played by species within a community, and nutrient cycling within a forest. (*Source*: Helms, J. (ed.). 1998. The Dictionary of Forestry. Society of American Foresters. USA).

[10]**Biodiversity**–The variety of life on Earth and the natural patterns it forms, including all species of life and the genes that each individual contains, as the critical inter-relationships or "ecosystems" which those species form. (*Source*: BIOTECanada. What is Biotechnology?Glossary.http://www.biotechca/EN/wht_glossary.html).

[11]**Biodiversity**–Biodiversity is the property of living systems of being distinct, that is, different, unlike. The word is a contraction of Biological Diversity, *i.e.* the diversity of living beings. Life comes in an almost infinite variety of fascinating and enchanting forms, from microsopically small unicellular species to giant whales and elephants. In turn, species are formed by different kinds of populations, these by different kinds of individuals, and these by different types of organs, tissues, cells, and genes. Diversity surround us, engulfs us, and not only in the living world. The inanimate world is also highly diverse. The rare, the peculiar, is to encounter living beings that are identical. [Source: Solbrig, O.T. et al. 1992: Biodiversity and Global Change. International Union of Biological Sciences (IUBS)].

B. BIOTECHNOLOGY

[1]**Biotechnology**–Any technological application that uses biological systems, living organisms, or derivatives thereof, to make or modify products or processes for specific use. (*Source*: The Convention on Biological Diversity. Article 2. UNEP 1992).

[2]**Biotechnology**–The application of scientific and engineering principles to the processing of materials by biological agents to provide goods and services. Biotechnology comprises any technique that uses living organisms to make or modify a product, to improve plants or animals or to develop micro-organisms for specific purposes. (*Source*: Schmidt, L., 1997). Tree Improvement Glossary. Technical Note Nr. 46. Danida Forest Seed Centre).

[3]**Biotechnology**:Any technique that uses living organisms (or part of organisms) to make or modify products, to improve plants or animals, or to develop micro-organisms for specific uses. It refers to the technological applications of biological processes, including a number of individual techniques such as recombinant DNA-molecule manipulation, protein engineering, cell fusion, nucleotide synthesis, monoclonal antibody use and production, product recovery, and

unique fermentation techniques, *e.g.* biocatalysis using immobilized enzymes. (*Source*:IBPGR (comp.), 1991. Elsevier's Dictionary of Plant Genetic Resources. Italy).

[4]**Biotechnology**–The collection of industrial processes that involve the use of biological systems. For some industries, these processes involve the use of genetically engineered micro-organisms. (*Source*: King, R.C.; Stansfield, W.D.; 1997. A Dictionary of Genetics. Oxford University Press).

[5]**Biotechnology**–The application of science and engineering in the direct or indirect use of living organisms, or parts of organisms, in their natural or modified forms. (*Source*: Canadian Forest Service. Science Branch. 1998. Biotech@nrcan.gc.ca).

[6]**Biotechnology**–The scientific manipulation of living organisms, especially at the molecular genetic level, to produce useful products. Gene splicing and use of recombinant DNA (rDNA) are major techniques used. (*Source*: Hagedorn, S. A. An Agricultural and Environmental Biotechnology Annotated Dictionary. http://gophisb.biochem.vt.edu/resources/glossary.htm).

[7]**Biotechnology**–The industrial use of living organisms or biological techniques developed through basic research. Biotechnology products include antibiotics, insulin, interferon, recombinant DNA, and techniques such as waste recycling. Much older forms of biotechnology include breadmaking, cheese-making and brewing wine and beer. (*Source*: Bio-Tech Resources. Indiana University. Life Science Dictionary. http://biotech.chem.indiana).

[8]**Biotechnology**–1. Any technique that uses living organisms, or parts of organisms, to make or modify products, to improve plants or animals or to develop micro-organisms for specific uses (OTA-U.S. Congress, 1988). 2. Dubbed as the technology of life by altering the genetic make-up of living organisms through tissue culture, cloning existing cells, DNA changing and genetic engineering (Pecs, K. 1993) 3. The application of scientific and engineering principles to the processing of materials by biological agents to provide goods and services (Reinjtjes *et al.*, 1993). 4. Assists the conservation of plant and animal genetic resources through: new methods for collecting and storing genes (as seed and tissue culture), detection and elimination of diseases in genebank collections; identification of useful genes; improvement techniques for long-term storage; and safer and more efficient distribution of germplasms to users (Harvesting Nature's Diversity). 5. The use of advance genetic techniques to construct novel microbial and plant strains and obtain site-directed mutants ot improve the quantity and quality of products (Mc Graw-Hill Dictionary of Scientific and Technical Terms). [*Source*: SRD Project. 1996. Glossary of Terms relevant to: Sustainable Agriculture, Seed Production and Handling Technology, Genetic Conservation, Agroforestry, Indigenous Knowledge Systems Development and Related Fields and Issues. Ottawa, Canada].

[9]**Biotechnology**–It can be broadly defined as "using living organisms or their products for commercial purposes." As such, biotechnology has been practiced by human society since the beginning of recorded history in such activities as baking bread, brewing alcoholic beverages, or breeding, food crops or domestic animals.

[10]**Biotechnology**–A narrower and more specific definition of biotechnology is "the commercial application of living organisms or their products, which involves the deliberate manipulation of their DNA molecules" .... This definition implies a set of laboratory techniques developed within the last 20 years that have been responsible for the tremendous scientific and commercial interest in biotechnology, the founding of many new companies, and the redirection of research efforts and financial resources among established companies and Universities. These laboratory techniques provide scientists with a spectacular vision of the design and function of living organisms, and provide technologists in many fields with the tools to implement exciting commercial applications. (*Source*: Iowa State University. Principles of Biotechnology. (Bio-1). http://www.nal.usda.gov/Bic/Education_res/iastate info/bio1.html

[11]**Biotechnology**–The use of current technologies such as DNA technologies for the modification and improvement of biological systems. (*Source*: BIOTEC Canada. What is Biotechnology? Glossary. http://www.biotechca/EN/what_glossary.html).

[12]**Biotechnology**–Development of products by a biological process. Production may be carried out by using intact organisms (*e.g.*, yeasts and bacteria) or by using natural substances (*e.g.*, enzymes) from organisms. (*Source*: Côté, M. (ed.) 2000. Dictionary of Forestry. Ordre des ingénieurs forestiers du Québec).

C. GENETIC CONSERVATION

[1]**Genetic conservation**–The collection, maintenance, storage and sustainable management of genetic resources aimed at ensuring their continued existence, evolution and availability for future generations.

[2]**Genetic conservation**–The collection, maintenance and preservation of intra–and inter specific variation, *e.g.* a representative sample of the genetic variation of a particular species. (*Source*: IBPGR. 1991. Elsevier's Dictionary of Plant Genetic Resources. Rome).

[3]**Genetic conservation**–1. The wise and sustainable management of natural resources with objectives being human survival and development (J. Burley, Abstracts). 2. The management of human use of genetic resources so that they may yield the greatest sustainable benefit to the present generation, while maintaining their potential to meet the needs and aspirations of future generations (Conservation of genetic resources in tropical forest management: Principles and Concepts). 3. Collection, main-

tenance, and preservation of all segments of germplasm in a crop species and its wild relatives (T.T. Chang, 1976). (*Source*: S R D Project, 1996. Glossary of Terms. Ottawa, Canada).

[4]**Genetic conservation**–All activities including, *e.g.* collecting, maintenance, storage, management, protection and regeneration, aimed at ensuring the continued existence, evolution and availability of genetic resources; *in situ* and *ex situ*. (*Source*: Koski, V. *et al.*, 1997. Technical guidelines for genetic conservation of Norway spruce (*Picea abies* (L.) Karst.). EUFORGEN. IPGRI).

[5]**Gene conservation**–Protecting and maintaining the genetic variation of a species in order to keep a genetic resource for future research and improvement. (*Source*: Schmidt, L. 1997. Tree Improvement Glossary. Technical Note 46. Danida Forest Seed Centre).

Gene conservation–A sound gene conservation means that the methods selected ensure that the objectives are fulfilled while taking the genetic knowledge into account. (*Source*: Mátyás, C. (ed.) 1997. IUFRO World Series Vol. 6.e).

[6]**Gene conservation**–Any action to maintain an individual gene (phenotpye) in a population. A term to be avoided in favor of the term gene pool (or genetic resource) conservation (Nienstaedt, 1980). [*Source*: USDA Forest Service. 1980. A Glossary of Terms for Forest Tree Improvement Workshop. Sacramento, California].

[7]**Gene pool conservation** ***(genetic resource conservation)***–Any action that insures the maintenance and wise use of the evolutionary potential of the population (Nienstaedt, 1980). [Source: USDA Forest Service. 1980. A Glossary of Terms for Forest Tree Improvement Workshop. Sacramento, California].

[8]**Conservation**–The management of human use of the biosphere so that it may yield the greatest sustainable benefit to present generations while maintaining its potential to meet the needs and aspirations of future generations: Thus conservation is positive, embracing preservation, maintenance, sustainable utilization, restoration, and enhancement of the natural environment. (*Source*: IUCN/WWF/UNEP.1981. Conservation Strategy).

[9]**Conservation**–Conservation of a resource is best defined as the actions and policies that assure its continued availability and existence. (*Source*: FAO. 1989. Plant Genetic Resources. Conceptual framework. FAO. Rome).

[10]**Conservation**–1. Protection of plant and animal habitat 2. The management of a renewable natural resource with the objective of sustaining its productivity in perpetuity while providing for human use compatible with sustainability of the resource–*note* for a forest this may include managed, periodic cutting and removal of trees followed by regeneration conservation 3. The process or means of achieving recovery of viable populations. (*Source*: Helms,

J. (ed.) 1998. The Dictionary of Forestry. Society of American Foresters).

[11]**Conservation**–1. The management or control of human use of resources (biotic and abiotic) and activities on the planet, in an attempt to restore, enhance, protect, and sustain the quality and quantity of a desired mix of species and ecosystem conditions and processes for present and future generations. 2. The process or means of achieving and maintaining conservation objectives. Conservation is explicitly concerned with the temporal distribution of use; that is, carefully considered use now so that some (more, less, or the same amount) will still exist for use later on. The use in question could be *consumptive* (harvesting of a resource) or *non-consumptive* (retention of ecological reserves). There are many varied definitions of conservation in different fields. Most encompass the notion of judicious use by humans over time, which in some instances may mean no use at all or use that serves to enhance, rather than deplete, resources. (*Source*: Dunster, J. and K. 1996. Dictionary of Natural Resource Management. CAB International).

[12]**Conservation**–In forestry, the wise use of natural renewable resources. A key idea for understanding "conservation" is "use" by people. (Source: Fletcher, R. *et al.* 1994. Glossary of Woodland Words. Oregon State University).

[13]**Conservation**–1. The management of human use of the biosphere so that it may yield the greatest sustainable benefit to present generations, while maintaining its potential to meet the needs and aspirations of future generations. Thus conservation is positive, embracing preservation, maintenance, sustainable utilization, restoration, and enhancement of the natural environment (McNeeley *et al.*; IUCN, 1980). However, conservation means different things to different people, and advocates of conservation, whether of landscapes or of species (*in situ*), vary in their attitudes to local people. Although some believe that local communities, their knowledge, and their traditional life-styles must have an important roles in conservation, others believe conservation requires the tight restriction of human activities (and sometimes even human presence) in the targeted area (D.A. Posey and G. Dutfield). 2. Wise or rational utilization and protection of resources (Canopy International Vol. 15, No. 1). 3. A means of ensuring that plant and animal resoures are available for use by present and future generations (Harvesting Nature's Diversity). 4. Those measures concerned with the preservation, restoration, beneficia-tion, maximization, reutilization, substitution, allocation and integration of natural resources (Mc Graw-Hill Dictionary of Scientific and Technical Terms). [*Source*: SRD Project, 1996. Glossary of Terms relevant to: Sustainable Agriculture, Seed Production and Handling Technology, Genetic Conservation, Agroforestry, Indigenous Knowledge Systems Development and Related Fields and Issues. Ottawa, Canada].

[14]**Conservation**–The human use of the biosphere so that it may yield the greatest sustainable benefit to present generations while maintaining its potential to meet the needs and aspirations of future generations, and includes the preservation, maintenance, sustainable utilisation, restoration and enhancement of the environment. Modern conservation theory incorporates the notion that what is to be conserved is not so much the physical state of an ecological system as the ecological processes by which that state is created and maintained (Australia, Resource Assessment Commission 1991). Conservation applies both to forests designated for harvesting and to forests designated as ecological reserves, although the management goals are different (Maini and Carlisle 1974). [*Source*: Arid, P.L. Conservation of the Sustainable Development of Forests worldwide: A Compendium of Concepts and Terms. The Forestry Chronicle Vol. 70, No. 6, 1994].

[15]**Conservation**–The practice that permits the perpetuation of a resource. (*Source*: Ayad, W.G. 1980. A Glossary of Plant Genetic Resources Terms. IBPGR, Rome).

[16]**Conservation**–Conservation is seen as the antithesis of use and in this kind of conflict often *conservation* and *breeding* are seen as mutually exclusive. As long as forests and their genes are assumed as fixed resources, both genetic and management issues will be solved along a one-dimensional axis of production versus conservation. On the other hand, if we view resources as multidimensional with many options for productivity and conservation, a more rational solution may be possible. (*Source*: Mátyás, C. (ed.) 1997. IUFRO World Series Vol. 6).

[17]**Conservation**–The management of human use of organisms or ecosystems to ensure such use is sustainable. Besides Sustainable Use, conservation includes Protection, Maintenance, Rehabilitation, Restoration, and Enhancement of populations and ecosystems. (*Source*: IUCN, UNEP, WWF, 1991: Caring for the Earth. A Strategy for Sustainable Living).

[18]**Conservation**–Judicious case and management of nature and natural resources for the benefit of human society and for ethical reasons. (*Source*: UNEP, 1995. Global Biodiversity Assessment. Annex. 6, Glossary).

D. EVOLUTIONARY PROCESS

The changes in genetic constitution of a population or group of populations in successive generations. Particularly long-term changes accompanying ecotype, race, subspecies, genus, and family formation.

The causative processes include mutation, recombination, drift, isolation, and natural selection. Evolution has no determined endpoint.

[1]**Evolution**–The step-wise development (or extinction) of biological groups, as a result of natural selection and increase of hereditary variants in the

population. A small proportion of individuals with a particular non-favourable genetic mark up may survive in a large population. Through sucessive generations, change in the environment or a new environment to which the organisms have migrated may favor the survival and reproduction of the aberrant individuals with a consequent increase of their genes in the population. The causative processes include mutation, recombination, drift, isolation, and natural selection. (*Source*: Schmidt, L. 1997. Tree Improvement Glossary. Technical Note 46. Danida Forest Seed Centre).

[2]**Evolution**–A cumulative genetic change in a population of organisms related by descent, over time. It is typically the result of natural selection but can also be due to random genetic drift. Evolution has no determined endpoint. (*Source*: Dunster, J.& K 1996. Dictionary of Natural Resource Management. CAB International).

Evolution–Long-time changes in gene frequency and phenotypic characteristics of a population or group of populations. (*Source:* Wright, J.W. 1976. Introduction to Forest Genetics. Michigan State University, New York).

Evolution–In biological terms: a change in the genetic composition of a population over time. (*Source*: BioTech Life Science Dictionary. Indiana University).

[3]**Evolution**–The changes in genetic constitution of a group in successive generations. Particularly long-time changes accompanying ecotype, race, subspecies, species, genus, and family formation. (*Source*: FAO. Genetics of Forest Tree Improvement).

[4]**Evolution**–The process of cumulative change in response to changing environmental conditions. The theory that species have changed in response to natural selection. (*Source*: Žumer-Linder, M. 1979. Environmental World-List; Swedish University of Agricultural Sciences. Uppsala).

[5]**Evolution**–1) The change in life over time by adaptation, variation, over-reproduction, and differential survival/reproduction, a process referred to by Charles Darwin and Alfred Wallace as natural selection.

2) Descent with modification. (Source: Online Biology Book: Glossary).

E. GENETIC POLLUTION

[1]**Genetic pollution**–Uncontrolled spread of genetic information (frequently referring to transgenes) into the genomes of organisms in which such genes are not present in nature. (*Source*: Zaid, A. *et al.*, 1999. Glossary of biotechnology and genetic engineering. FAO Research and Technology Paper No. 7. ISBN 92-5-104369-8 http://www.fao.org/DOCREP/003/X3910E/X3910E00.htm).

[2]**Genetic pollution**–Uncontrolled escape of genetic information (frequently referring to products of genetic engineering) into the genomes of organisms in the environment where those genes never existed

before. (*Source*: Searchable Biotechnology Dictionary. University of Minnesota. http://www.plpa.agri.umn.edu/scag1500/definitions.html).

F. GENETIC DIVERSITY

[1]**Genetic diversity**–The genetic variability within a population or a species, usually assessed at three levels: (a) within breeding populations, (b) between breeding populations, and (c) within species. (*Source:* Helms, J. (ed.) 1998. The Dictionary of Forestry, Society of American Foresters).

[2]**Genetic diversity**–The genetic variability within a population or a species. The number and relative abundance of alleles. This is the foundation of all diversity, ad loss of genetic diversity within species is increasingly recognized as an important and largely undocumented problem, at least as serious as loss of entire species. Genetic diversity can be assessed at three levels: (1) diversity within breeding populations; (2) diversity between breeding populations within any one geographic area; and (3) diversity within the species. (*Source:* Dunster, J.& K. 1996. Dictionary of Natural Resource Management. CAB International).

[3]**Genetic diversity**–The condition of being genetically different and variable. (*Source:* Ayad, W.G. 1980. A Glossary of Plant Genetic Resources Terms. IBPGR Secretariat. AGP: IBPGR/80/11).

[4]**Genetic diversity**–It refers to the range of variability (for specific characters) among representative plants or seeds of a genetically pure seedlot. It describes the homogeneity or heterogeneity of seeds and ultimate plants composing a variety. On a broader scale, the term may also be used to describe the diversity of varieties growing in a given area Variation or differences within the genes of an offspring; variations can be visible, such as color, height or shape, or invisible like taste, resistance to pests and diseases. Makes it possible to produce new breeds of crops or animals and allow them to adapt to changing conditions (soils, climate, different management practices or agricultural uses) (Pecs, 1993). Existing condition of being genetically different (T.T. Chang, 1976). The total range of genetic differences displayed by plants of the same species (S. Ashworth, 1991). [*Source:* SRD Project, 1996. Glossary of terms relevant to Sustainable Agriculture, Seed Production and Handling Technology, Genetic Conservation, Agroforestry, Indigenous Knowledge Systems Development and Related Fields and Issues. Canada].

[5]**Genetic diversity**–It refers to the variation of genes within species. This covers distinct populations of the same species (such as the thousand of traditional rice varieties in India) or genetic variation within a population (which is very high among Indian rhinos, for example, very low among cheetahs). Until recently, measurements of genetic diversity were applied mainly to domesticated species and populations held in zoos or botanic

gardens, but increasingly the techniques are being applied to wild species. (*Source:* WRI. "What is Biological Diversity". In: "The Diversity of Life" http://w,ww.wri.org/biodiv.html).

[6]**Genetic diversity**–General concept: the amount of genotypic variability in a population. Quantitative definition: the number of different alleles per loci and the proportion of loci with more than one allele in a species or population. (*Source:* Maynard, C., 1996: Forest Genetics Glossary. SUNY College of Environmental Science and Forestry. http://www.esf.edu/course/cmaynard/GENE_GLOSSARY.html).

[7]**Genetic diversity**–It is one aspect of biological diversity. Genetic diversity occurs at gene level (the molecular level), the individual level, the population level, the species level, and the ecosystem level. (*Source:* FAO, 1989. Plant Genetic Resources. Conceptual framework).

[8]**Genetic diversity**–The variation in the genetic composition of individuals within or among species, varieties or breeds; the heritable genetic variation within and among populations (World Resources Institute et al. 1992). [*Source:* Aird, P. L. 1994. A Compendium of Concepts and Terms. The Forestry Chronicle Vol. 70, No. 6].

[9]**Genetic diversity**–It refers to the variety of genetic information contained in all of the individual plants, animals and micro-organisms. Genetic diversity occurs within and between populations of species as well as between species.It refers to the variation of genes within species. This covers genetic variation between distinct populations of the same species, such as the four varieties of white-cheeked rosella, Platycercus eximius. It also covers genetic variation within a population, which tends to be relatively high in widespread eucalyptus such as *Ecualyptus cloeiana, E. delegatensis,* and *E. saligna*. Genetic diversity can be measured using a variety of DNA-based and other techniques. (*Source:* Biodiversity and its value. Biodiversity Unit, Department of the Environment, Sport and Territories. Australia).

[10]**Genetic diversity**–The variety and frequency of different genes and/or genetic stocks. (*Source:* IUCN, UNEP, WWF; 1991: Caring for the Earth. A Strategy for Sustainable Living).

[11]**Genetic diversity**–In ecology, the number of species or other taxa in a particular ecological unit. (*Source:* King, R.C.; Stansfield, W.D.; 1997. A Dictionary of Genetics, Fifth edition, Oxford University Press).

[12]**Genetic diversity**–The range of a genepool; the amount of genetic variation present in a population or species as a consequence of its evolutionary pathways. (*Source:* IBPGR (comp.). 1991. Elsevier's Dictionary of Plant Genetic Resources. Rome).

[13]**Genetic diversity**–The measure of genetic variation present in a population as a consequence of its evolution. (*Source:* Koski, V. *et al.* 1997. EUFORGEN. IPGRI).

[14]**Genetic diversity**–The formation of individuals differing in genotype, or the presence of genotypically different individuals, in contrast to environmentally induced differences which, as a rule, cause only temporary, nonheritable changes to the phenotype (Rieger et al., 1968) These terms are also used to imply the maintenance of the gene pool. (*Source:* USDA Forest Service. 1980. A Glossary of Terms for Forest Tree Improvement Workshop. Sacramento, California).

G. GENETIC VARIATION

The occurrence of genetic variants (alleles, genes or genotypes).

Genetic variation is brought about by a change in genes, as distinct from differences due to environmental factors. (Source: Koski, V. *et al.*, 1997, EUFORGEN, IPGRI).

[1]**Genetic variation**–Genetic diversity or variation is not explicitly defined in the Convention. Genetic variation includes genetic differences between species and within species. The genetic diversity of a species can be divided into inter-population diversity and intra-population diversity, and further into the diversity within an individual expressed by differences between alleles in the two chromosomes of diploid organisms (degree of individual heterozygosity). [Source: Graudal, L. et al. 1997. Technical Note 48. Danida Forest Seed Centre].

[2]**Genetic variation**–Variation due to the contribution of segregating genes and gene interactions. (Source: Ayad, W.G. 1980. A Glossary of Plant Genetic Resources IBPGR Secretariat. AGP: IBPGR/80/11).

[3]**Genetic variation**–The occurrence of differences among individuals of the same species attributable to differences in their genetic composition or the environment in which they were raised–note quantitative differences in a given trait are assessed by their variance. [Source: Helms, J. (ed.) 1998. The Dictionary of Forestry, Society of American Foresters (SAF)].

[4]**Genetic variation**–Differences in form or function between individuals or populations or species. Variations may or may not be heritable and are caused by both genetic and environmental factors. Natural variation is that which occurs spontaneously. (Source: IBPGR (comp.) 1991. Elsevier's Dictionary of Plant Genetic Resources, Rome).

[5]**Genetic variation**–Differences in the frequency of gene and trait among individual organisms within a population. (Source: Hagedorn, S. A. An Annotated Dictionary. http://gophisb.biochem.vt.edu/resources/glossary.htm).

[6]**Genetic variation**–Difference in performance and characters of individuals due to internal and external factors. In a natural population the phenotypic variation is the product of developmental, environmental and genetic variation. In trials, sampling variation and experimental error make up the residual variation. (Source: Schmidt, L. 1997. Tree Improvement Glossary. Technical Note 46, Danida Forest Seed Centre).

[7]**Geographic variation**–The phenotypic differences among native trees growing in different portions of a species' range. If the differences are

largely genetic rather than environmental, the variation is usually specified as racial, ecotypical, or clinal. (*Source*: Schmidt, L. Glossary of Terms Used in Forest Tree Improvement. UNDP/ FAO (RAS/91/004)).

8**Phenotypic variation**–The total biological variation of a given character. (*Source*: Ayad, W.G. 1980. Glossary of Plant Genetic Resources Terms. IBPGR Secretariat. AGP:IBPGR/80/11).

H. GENETIC RESOURCES

[1]**Genetic resources**–Genetic material of actual or potential value. (*Source:* The Convention on Biological Diversity. Article 2. UNEP 1992).

[2]**Genetic resources**–Genetic resources are the heritable characteristics of a plant or animal of real or potential benefit to people. The term includes modern and traditional cultivars and breeds; special genetic stocks (breeding lines, mutants, etc); wild relatives of domesticated species; and genetic variants of wild resource species. A ,wild genetic resource' is the wild relative of a plant or animal that is already known to be of economic importance. The reasons for conserving such a resource include the provision of direct and indirect economic benefits. However, the conserved genetic material must be made available to the people who require it to improve the productivity, quality, or pest resistance of utilized plants or animals. (*Source*: Dunster, J. and K. 1996. Dictionary of Natural Resource Management. CAB International).

[3]**Genetic resources**–Genetic resources are the genes, stored as germplasm (seeds, tubers or other reproductive parts of plants), that can be used to develop new crops and crop varieties or to protect existing crops from pests, diseases or environmental stresses. (*Source*: American Genetic Resources Alliance, 1998. What are Genetic Resources? http://www.amgra.org/grbkgrd.htm).

[4]**Genetic resources**–1. In a strict sense, the physical germplasm (hereditary materials) which carries the genetic characteristics of life forms. In a broad sense, technologies and social and environmental systems through which germplasm is a cash-economic resources (Cooper *et al.*, 1992). 2. Plant and animal stock with distinct inheritable characteristics of (potential) use within an agroecosystem (Reinjtjes *et al.*, 1993). 4. Germplasm that includes the entire array of cultivars in the crop species, related wild species in the genus, and hybrids between the wild and cultivated species (T.T. Chang, 1976). [*Source*: SRD Project. 1996. Glossary of Terms relevant to: relevant to: Sustainable Agriculture, Seed Production and Handling Technology, Genetic Conservation, Agroforestry, Indigenous Knowledge Systems Development and Related Fields and Issues, Canada].

[5]**Genetic resources**–Germplasm of plants, animals, or other organisms, containing useful characters of actual or potential value. In a domesticated species, it is the sum of all the genetic combinations produced in the process of

evolution. [*Source*: IBPGR (comp.). 1991. Elsevier's Dictionary of Plant Genetic Resources. Rome].

[6]**Genetic resources**–The economic, scientific or social value of the heritable materials contained within and between species. (*Source*: Kemp, R.H. 1993. Principles and concepts. FAO Forestry Paper 107).

[7]**Genetic resources**–In the same way that the term forest resources refers to te usefulness of the forests for the production of timber or other products for human benefit, the term genetic resources implies that elements of the genetic variability of the trees and other plants and animals will be used to meet human needs and objectives. The other important aspect of the genetic resources of natural forests, especially the tropical forests, is their great diversity, and this range of variation provides the basis for selection and improvement of the products and other benefits to meet future needs, so far as they can be foreseen. (*Source*: Kemp, R.H. Principles and concepts. FAO Forestry Paper 107).

8**Genetic resources**–When genetic diversity can be used for plant domestication and improvement, diversity found in the original plant material is considered a genetic resource. Plant genetic resources thus refers to the economic, scientific or societal value of the heritable materials contained within and among species. (*Source*: FAO. 1989. Conceptual framework).

[9]**Gene resource (genetic resource)**–The total genetic information encoded in the sum total of the genes in all the populations of a species. Thus it is the sum of the genetic information in the gene pools of the species. Where interbreeding species overlap the species complex may represent the gene resource (Nienstaedt, 1980). [*Source*: USDA Forest Service. 1980. A Glossary of Terms for Forest Tree Improvement Workshop. Sacramento, California].

[10]**Genetic resources**–A unit of heritable variability of actual or potential value. (*Source*: Schmidt, L. 1997. Tree Improvement Glossary. Technical Note 46. Danida Forest Seed Centre).

I. GENETIC RESOURCE MANAGEMENT

[1]**Management of genetic resources**–The management of forest genetic resources to ensure at the same time their conservation, improvement and sustainable use is a complex challenge. Fortunately, when simple basic principles are applied, the production of goods and services is generally compatible with the genetic conservation and development of a given forest tree species. (*Source*: FA Forestry. Conservation and Management of Forest Genetic Resources. http://www.fao.org/FORESTRY/FOR/FORM/FOGENRES/homepage/Insitu-e.stm).

[2]**Management of genetic resources**–Higher plant, microbial and insect germplasm should be acquired and safeguarded through the expansion of genebanks or *in situ* preserves for long-term accessibility. The genetic content of acquired germplasm should be characterized to insure

broad-spectrum genetic variability while minimizing genetic redundancy. The agricultural potential of unimproved germplasm must be assessed. (*Source*: USDA Agricultural Research Service. Plant, Microbial, and Insect Genetic Resources, Genomics and Genetic Improvement. Program Component Definitions. http://www.nps.ars.usda.gov/programs/programs.htm?npnumber=301&docid=791).

[3]**Management of genetic resources**–Effective conservation and management of genetic resources in agriculture are key factors in achieving long-term and sustainable food security. They are highly important for poverty reduction, since an estimated 85–90 per cent of the worlds poor rely on biological products to meet their basic needs, *i.e.* food, fuel, medicine, shelter and transportation. Additionally, in the wake of globalisation and increased privatisation, agricultural genetic resources have become a highly political issue and in some cases this poses a serious threat to their continued use for the improved well being of poor people. (*Source*: http://ecart.iao.florence.it/capstat/grmfr.htm).

J. GENETIC MAKEUP OR GENOME

All the genetic material in the chromosomes of a particular organism.

The genetic material inherited from either parent.

[1]**Genome**–All the genetic material in the chromosomes of a particular organism; its size is generally given as its total number of base pairs. (*Source*: Bio-Tech Life Science Dictionary, Indiana University).

[2]**Genome**–All the genes of a living organism. Its complete set of chromosomes (DNA), with their associated genes. [*Source*: Forest Genetics Council. BC, Canada. http://www.fgcouncil/.bc.ca/framdocs.htm].

[3]**Genome**–1 the genetic complement of an individual. 2 all of the DNA sequences in a single (haploid) set of chromosomes. The genetic material inherited from either parent. [*Source*: Dunster, J.& K. 1996. Dictionary of Natural Resource Management. CAB International].

[4]**Genome**–1. The set of genes carried by an individual. 2. The set of genes shared by members of a reproductive unit such as a population or species. [*Source*: On-Line Biology Book: Glossary].

[5]**Genome**–The total complement of genes contained in a cell or virus; commonly used to refer to all genes present in 1 complete set of chromosomes in eukaryotes. [*Source*: Klopfenstein, N. B. *et al.* (ed.) 1997. USDA Forest Service General Technical Report RM-GTR-297].

[6]**Genome**–A complete haploid set of chromosomes as found in a gamete. [*Source*: Helms, J. (ed.) 1997. The Dictionary of Forestry. Society of American Foresters].

[7]**Genome**–The genetic content of a cell or virus; in eukaryotes, it sometimes refers to only one complete (haploid) chromosome set. [*Source*: Ayala, F.J./

Kiger, J.A.; Modern Genetics, University of California, Davis].

[8]**Genetic makeup**–Total genetic content of an individual or species (also called genome). [*Source*: Schmidt; L. 1997. Tree Improvement Glossary. Technical Note 46. Danida Forest Seed Centre].

K. GENOTYPE

The genetic constitution of an organism as distinguished from its appearance or phentoype.

Also the gene classification of this constitution expressed in a formula.

[1]**Genotype**–(1) An individual's hereditary constitution, with or without phenotypic expression of the one or more charcaters it underlies. Also the gene classification of this constitution expressed in a formula. The genotype is determined chiefly from performance of progeny and other relatives. It interacts with the environment to produce the phenotype. (2) Individual(s) characterized by a certain genetic constitution. [*Source*: Schmidt, L. 1997. Tree Improvement Glossary. Technical Note 46. Danida Forest Seed Centre].

[2]**Genotype**–The genetic make up of an organism, this being the sum total of all the genetic information in the organism. In analysis of the genetic constitution of a few gene loci, the genotype is all the characteristics on the chromosome, even if they are not expressed in the phenotype. [*Source*: Dunster, J and K. 1996. Dictionary of Natural Resource Management. CAB International].

[3]**Genotype**–Genetic constitution of an individual tree possessing a particular set of alleles (*i.e.* different forms of genes which may occupy the same position on a chromosome). [*Source*: Koski, V. *et al.* Technical Guidelines for genetic conservation of Norway spruce (*Picea abies* (L.) Karst.). Euforgen. IPGRI.].

[4]**Genotype**–(1) The entire genetic constitution, expressed or latent, of an organism. (2) The genetic constitution of an individual with respect to a few genes under consideration. (3) A group of organisms having similar genetic constitution. [*Source*: Wright, J. W. 1976. Introduction to Forest Genetics. Michigan State University, New York].

[5]**Genotype**–The genetic (alleleic) makeup of an organism with regard to an observed traits. [Source: On-Line Biology Book: Glossary].

[6]**Genotype**–Genetic constitution as contrasted wit the characteristics manifested by the organism (phenotype).[Source: MacMillan, S. (comp.) 1980. Penguin Dictionary of Biology].

[7]**Genotype**–(1) An individual's hereditary constitution, with or without penotypic expression of the one or more characters it underlies. The genotype is determined chiefly from performance of progeny and other relatives. It interacts with the environment to produce the phenotype. (2) Individuals characterised by a certain genic constitution. [*Source*: FAO. A guide to forest seed handling. Forestry 20/2].

[8]**Genotype**–The genetical potential of the tree when environmental factors are excluded, *i.e.* only determined by the genes of the tree. [*Source*: ITTO. Regional Centre for Forest Management Malaysia].

[9]**Genotype**–An individuals entire genetic or hereditary constitution, with or without phenotypic expression. The genotype interacts with the environment to produce the phenotype. [*Source*: ODA. Silvicultural Manual for the Solomon Islands, Solomon Island Forest Record N°6, ODA Forestry Series N°1].

[10]**Genotype**–The sum total of the genetic information contained in an organism with respect to one or a few gene loci under consideration (cf. phenotype). [*Source*: Ayala, F.J./ Kiger, J.A.; Modern Genetics, University of California, Davis].

[11]**Genotype**–The specific set of genes possessed by an individual, both expressed and recessive. [*Source*: Maynard, C. 1996 Forest Genetics Glossary].

[12]**Genotype** (adj. genotypic)–The genetic constitution of an organism. Compare phenotype. [*Source*: Bio-Tech Life Science Dictionary, Indiana University].

[13]**Genotype**–This term may be used in a limited sense to describe the genetic constitution of an individual in terms of a few specific genes, or in a general sense to include the entire genetic constitution (expressed or latent) of an individual. [*Source*: FAO. Genetics of Forest Tree Improvement].

[14]**Genotype**–The genetic constitution of an organism or virus as distinguished from its appearance or phenotype; the allelic composition of one or more genes of interest. [*Source*: Klopfenstein, N.B. *et al*. 1997. USDA Forest Service General Technical Report RM-GTR-297].

L.GENE POOL

The total genetic information present in a breeding population or species at one time.

More often applied as a common term for all genes present in plant or animal populations from specific regions, *e.g.* Amazonian forests.

[1]**Gene pool**–The total genetic information possessed by the reproductive members of a population of sexually reproducing organisms. [*Source:* FAO, UNEP. The methodology of conservation of forest genetic resources. Report on a pilot study].

[2]**Gene pool**–The total sum of all the genes and their alleles present in a breeding population or species at one time. [*Source:* Schmidt, L. 1997. Tree Improvement Glossary. Technical Note 46. Danida Forest Seed Centre].

[3]**Gene pool**–The total of all the genes of all breeding individuals available within a population at any one time. [*Source:* Dunster, J. and K. 1996. Dictionary of Natural Resource Management. CAB International].

[4]**Gene pool**–All of the genes in a species, sub-species or interbreeding group of organisms. [*Source:* Côté, M. (éd.) 2000. Dictionary of Forestry. Ordre des Ingénieurs forestiers du Québec. Canada].

[5]**Gene pool**–Total number of genes/gene complexes in an organism's population. [*Source:* MacMillan, S. 1980. Penguin Dictionary of Biology].

[6]**Gene pool**–The total of all the genes of a species. More often applied as a common term for all genes present in plant or animal populations from specific regions, *e.g.* Amazonian forests. [*Source:* Żumer-Linder, M. 1979. Ecological studies 3. Swedish University of Agricultural Sciences, Uppsala].

[7]**Gene pool**–The sum of all genetic information encoded in genes and their alternative forms (alleles) present in a population at a given time. [*Source:* Koski, V. Technical guidelines for conservation of Norway Spruce. Euforgen. IPGRI].

[8]**Gene pool**–The sum total of all the genetic variation in the breeding population of a species and closely related species capable of crossing with the species.

Gene pool, primary–The sum total of all the genetic variation in the breeding populations of a species and closely related species that commonly interbreed with, or can be routinely crossed with, the species.

Gene pool, secondary–The sum total of all the genetic variation in the breeding populations of related species that can be crossed with the species using mentor pollen, embryo rescue or other unusual measures.

Gene pool, tertiary–The sum total of all the genetic variation in other organisms that cannot be crossed with the species. With the development of genetic engineering, it is theoretically possible to transfer genes isolated from any organism (plant, animals, virus, or bacterium) into a plant. This makes the line between the secondary and tertiary gene pools somewhat fuzzy. [*Source:* Maynard, C. 1996. Forest Genetics Glossary].

M. GERMPLASM

The genetic material which forms the physical basis of heredity and which is transmitted from one generation to the next by means of the germ cells.

Often synonymous with genetic material, when applied to plants it is the name given to seed or other material from which plants are propagated.

[1]**Germplasm**–Within an individual or group, the collective hereditary material that is the physical basis for inheritance; *i.e.* the genotype, with particular reference to its transmission to the next generation. [*Source*: Schmidt, L., 1997. Tree Improvement Glossary. Technical Note 46. Danida Forest Seed Centre].

[2]**Germplasm**–1) The genetic material which forms the physical basis of heredity and which is transmitted from one generation to the next by means of the germ cells. 2) An individual or clone representing a type, species or culture, that may be held in a repository for agronomic, historic or other reasons. [*Source*: IBPGR (comp.) 1991. Elsevier's Dictionary of Plant Genetic Resources].

[3]**Germplasm**–The genetic material of an individual organism contained in

the seed, pollens, sperms, eggs, or embryos, which constitutes the heritable characteristics of the organism. The plasm is stored in carefully controlled conditions for future breeding, genetic engineering, derivation of pharmacological products, or conservation of species. Germ plasm can be stored in seed banks, sperm banks, or gene banks. [*Source*: Dunster, J. and K. 1996. Dictionary of Natural Resource Management. CAB International].

[4]**Germplasm**–The sum total of the genetic material in a plant: crop plants plus primitive cultivars, landraces, and wild and weedy relatives; also referred to as the wild species, genes from the wild, the world's gene pool of the plants, the genetic largess, the common heritage of mankind, plant genetic resources, the "public property" of the Third World countries/industrial nations, the "primitive landraces", of the south or the "elite commercial cultivar" of the north (Pecs, 1993).

Living substance of the cell nucleus that determines the hereditary properties of organisms and that transmits these properties from makeup of organism (O.L. Justice and L.N. Bass, 1978). 6. Any plant genetic material used for plant propagation and breeding, with emphasis on its genetic contents (N.P. Louwaars, 1994).

The genetic material, especially its specific molecular and chemical constitution, that comprises the physical basis of the inherited qualities of an organism (W.V. Reid and K.R: Miller, 1993). 8. Often synonymous with "genetic material", when applied to plants it is the name given to seed or other material from which plants are propagated (IDRC 1985; D.A. Posey and G. Dutfield, 1996). [*Source*: SRD Project. 1996. Glossary of Terms relevant to: Sustainable Agriculture, Seed Production and Handling Technology, Genetic Conservation, Agroforestry, Indigenous Knowledge Systems Development and Related Fields and Issues. Ottawa].

[5]**Germplasm**–controls heritable traits of species represented in the gene pool. [*Source*: FAO. 1989. Plant Genetic Resources. Conceptual framework].

[6]**Genetic material**–It means any material of plant, animal, microbial or other origin containing functions of heredity. [*Source*: UNEP. Convention on Biological Diversity. Article 2.June 1992].

[7]**Germ plasm = Germplasm**; genetic make-up of a population or individual organism; often used interchangeably with "gene pool". [*Source*: Dermine, P. 1990. Vocabulary of Agriculture. Terminology Bulletin 197. Minister of Supply and Services Canada].

N. PHENOTYPE

The observable appearance of an organism, as determined by environmental and genetic influences (in contrast to genotype). [*Source*: Glossary of Biodiversity Terms. UNEP-WCMC].

[1]**Phenotype**–The plant or character as we see it; state, description, or degree of expression of a characters; the product of the interaction of the

genes of an organism (genotype) with the environment. When the total character expressions of an individual are considered, the phenotype describes the individual. Similar phenotypes do not necessarily breed alike. [*Source*: Schmidt, L. 1997. Tree Improvement Glossary. Technical Note 46. Danida Forest Seed Centre].

[2]**Phenotype**–The morphological, physiological, biochemical, behavioral, and other properties of an organism that develop through the interaction of genes and environment. (see genotype). [*Source*: Biodiversity Glossary of Terms. World Resources Institute. WRI. http://www.wri.org/biodiv/gbs-glos.html].

[3]**Phenotype**–(Gr. *phaneros*, showing + type) The visible appearance or set of traits of an organism resulting from the combined action of genotype and environment. *cf* genotype. [*Source*: Zaid, A. *et al.*, 1999. Glossary of biotechnology and genetic engineering. FAO Research and Technology Paper No. 7. ISBN 92-5-104369-8 http://www.fao.org/DOCREP/003/X3910E/X3910E00.htm].

[4]**Phenotype**–An organism as observed, *i.e.* as judged by its visually perceptible characters resulting from the interaction of its genotype with the environment. Note: Identical phenotypes do not necessarily breed alike. Cf. genotype, heritability. [*Source*: Ford-Robertson, F.C. Terminology of Forest Science, Technology Practice and Products. Society of American Foresters, FAO/IUFRO Committee on Forestry Bibliography and Terminology. 1971].

O. PROVENANCE

The original geographic source of seed, pollen, or propagules.

In forestry literature the term is usually considered synonymous with "geographic origin", and preferred to "origin".

Provenance–(1) The original geographic source of seed, pollen, or propagules. In forest tree breeding the term usually refers to the original native sources of a population. When a population is removed from its sources and has grown elsewhere for a number of generations, it is referred to as a 'land race'. (2) The place in which any stand of trees are growing. The stand may be indigenous or non-indigenous. [*Source*: Schmidt, L. Danida. 1997. Tree Improvement Glossary. Technical Note 46. Danida Forest Seed Centre].

[1]**Provenance**–The region or geographical source where a plant or animal was originally found and is native, and where its genetic constitution has developed through natural selection in between periods of glaciation. [*Source*: Dunster, J.&K. 1996. Dictionary of Natural Resource Management. CAB International].

[2]**Provenance**–For seeding material, the provenance is the harvest location; for plants, it is both the harvest location and the location of the nursery.[Source:The Habitat Restoration Group, BioTech Life Science Dictionary, BioTech Resources and Indiana University].

[3]**Provenance**–It is the geographic location of a seed source. [*Source*: ITTO. Technical Guidelines for the

Establishment and Management of *Ex situ* Conservation Stands of Tropical Timber Species].

[4]**Provenance**–The geographic source or location to which plants are native and within which their genetic characteristics have been developed through natural selection, *e.g.* tree provenances. [*Source*: Zumer-Linder, M. 1979. Environmental World List; Swedish University of Agricultural Sciences, Uppsala].

[5]**Provenance**–The geographic origin of a population. Mostly (but not always in European usage) the ultimate natural origin, implying where the population evolved prior to human intervention. [*Source*: http://www.fgcouncil.Bc.ca/framdocs.htm].

[6]**Provenance**–The original geographic source of seed, pollen, or propagules. [*Source*: FAO. UNEP. The methodology of conservation of forest genetic resources. Report on a pilot study].

[7]**Provenance**–The original geographic source of seed or pollen (*also see* landrace and derived provenance), or the place in which any stand of trees is growing. The stand may be indigenous or exotic. [*Source*: ODA. Silvicultural Manual for the Solomon Islands, Solomon Island Forest Record N°6, ODA Forestry Series No.1].

[8]**Provenance**–The place in which any stand of trees is growing. The stand may be autochthonous or non-autochthonous (see–> origin). [*Source*: Koski, V. *et al.* Technical guidelines for genetic conservation. Eurforgen. IPGRI].

[9]**Provenance**–The ultimate natural origin of a tree or group of trees. In forestry literature the term is usually considered synonymous with "geographic origin", and preferred to "origin", which could mean "nursery of origin", "seedhouse of origin", or "method of propagation". Sometimes used to denote the trees having a given place of origin. [*Source*: Wright, J. 1976. Introduction to Forest Genetics, Department of Forestry Michigan State University. Academic Press. (AP) New York].

[10]**Provenance**–Fr. (provenience, L.) in European usage, the place where seeds were collected; in North America, synonymous with seed source [*Source*: SAETTEM].

P. SPECIES

A population or series of populations of organisms that are capable of interbreeding freely with each other but not with members of other species. [Source:Keystone Center 1991].

[1]**Species**–A population of organisms which are able to interbreed freely under natural conditions. A species represents a group of organisms which has evolved distinct inheritable features and occupies unique geographical area. Species do not usually interbreed freely with other species. [*Source*: Schmidt, L., 1997. Tree Improvement Glossary. Technical Note 46. Danida Forest Seed Centre].

[2]**Species**–A group of individual that have their major characteristics in common and (usually) can only breed with each other. [*Source*:

Dunster, J. and K. 1996. Dictionary of Natural Resource Management. CAB International].

[3]**Species**–Groups of populations (which are groups of individuals living together that are separated from other such groups) which can potentially interbreed or are actually interbreeding, that can successfully produce viable, fertile offspring (without the help of human technology). [*Source*: Mayr, E. 1969. In: BioTech Life Sience Dictionary, BioTech Resources and Indiana University].

[4]**Species**–It is a group of individuals of similar morphology that are able to breed with each other but not with individuals from outside this group. [*Source*: ITTO. Technical Guidelines for the Establishment and Management of *Ex situ* Conservation Stands of Tropical Timber Species].

[5]**Species**–One or more populations of interbreeding or potentially interbreeding organism that are reproductively isolated in nature from all other organisms. Populations of individuals capable of interbreeding and producing viable, fertile offspring. The least inclusive taxonomic category commonly used. [*Source*: On-Line Biology Book Glossary].

[6]**Species**–One or more populations, the individuals of which can interbreed, but which in nature cannot exchange genes with members belonging to other species. The main category of taxonomic classification. [*Source*: Burley, J. Wood, P.J. (comp.) 1976. A manual on species and provenance research with particular reference to the tropics. Tropical Forestry Papers No. 10, CFI, University of Oxford].

[7]**Species**–The unit of taxonomic classification in which genera are sub-divided. A group of similar individuals different from other similar arrays of individuals. In sexually reproducing organisms, the maximum inter breeding group isolated from other species by barriers of sterility or reproductive capacity. [*Source*: FAO/UNEP. The methodology of conservation of forest genetic resources. Report on a pilot study].

[8]**Species**–A group of similar organisms, capable of interbreeding, and more or less distinctly different in geographic range and/or morphological characteristics from other species in the same genus. [*Source*: Wright, J.W. 1976. Introduction to Forest Genetics. Department of Forestry Michigan State University. East Lansing, Michigan. Academic Press. (AP) New York].

[9]**Species**–(ecol) The main category of taxonomic classification into which genera are subdivided, comprising a group of similar interbreeding individuals sharing a common morphology, physiology, and reproductive process. [*Source*: Côté, M. (éd.) 2000. Dictionary of Forestry. Ordre des ingénieurs forestiers du Québec. Les Presses de l'Université Laval. Québec, Canada].

[10]**Species**–(silv) A group of individuals that have their major characteristics in common, *e.g.* tree species. [*Source*: Côté, M. (éd.) 2000. Dictionary of

Forestry. Ordre des ingénieurs forestiers du Québec. Les Presses de l'Université Laval. Québec, Canada].

[11]**Species**–Category of taxonomic classification below genus rank including individuals with similar morphological characteristics and defined by breeding potential and gene flow. Interbreeding occurs between individuals within a species resulting in gene flow to the next generation. Such interbreeding does not normally occur between individuals of different species. [*Source*: SAETTEM].

[12]**Species**–The smallest unit of classification commonly used. Groups of interbreeding natural populations which are reproductively isolated from other such groups. [*Source*: Žumer-Linder, M. 1979. Environmental World-List. Swedish University of Agricultural Sciences, Ecological studies 3, Uppsala].

References

Aird, P. L. 1994. A Compendium of Concepts and Terms, The Forestry Chronicle Vol. 70, No. 6.

Aird, P. L. Conservation of the Sustainable Development of Forests worldwide: A Compendium of Concepts and Terms. The Forestry Chronicle Vol. 70, No. 6, 1994.

American Genetic Resources Alliance, 1998. What are Genetic Resources? http://www.amgra.org/grbkgrd.htm.

Ayad, W.G. 1980. A Glossary of Plant Genetic Resources Terms. IBPGR Secretariat. AGP: IBPGR/80/11.

Ayala, F.J./Kiger, J. A.; Modern Genetics, University of California, Davis.

Biodiversity and its value. Biodiversity Unit. Department of the Environment, Sport and Territories. Australia.

Biodiversity Glossary of Terms. World Resources Institute. WRI. http://www.wri.org/biodiv/gbs-glos.html.

BIOTECanada. What is Biotechnology? Glossary. http://www.biotechca/EN/what_glossary.html.

Bio Tech Life Science Dictionary. Indiana University.

Burley, J. Wood, P.J. (comp.) 1976. A manual on species and provenance research with particular reference to the tropics. Tropical Forestry Papers No. 10, CFI, University of Oxford.

Canadian Forest Service, Science Branch. 1998. Biotech@nrcan.gc.ca.

Côté, M. (éd.) 2000. Dictionary of Forestry. Ordre des Ingénieurs forestiers du Québec: Canada.

Dermine, P. 1990. Vocabulary of Agriculture, Terminology Bulletin 197. Minister of Supply and Services Canada.

Dunster, J. and K. 1996. Dictionary of Natural Resource Management. CAB International.

Durrenberger, Robert W. 1973. *Dictionary of the Environmental Sciences*. Palo Alta, Ca.: National Press Books.

FA Forestry. Conservation and Management of Forest Genetic Resources. http://www.fao.org/FORESTRY/FOR/FORM/FOGENRES/homepage/Insitu-e.stm.

FAO. 1989. Plant Genetic Resources. Conceptual framework. FAO. Rome.

FAO. A guide to forest seed handling. Forestry 20/2.

FAO. Genetics of Forest Tree Improvement.

FAO. UNEP. The methodology of conservation of forest genetic resources. Report on a pilot study.

Fletcher, R. *et al.*, 1994. Glossary of Woodland Words. Oregon State University.

Ford-Robertson, F.C. Terminology of Forest Science, Technology Practice and Products. Society of American Foresters, FAO/IUFRO Committee on Forestry Bibliography and Terminology. 1971.

Forest Genetics Council. BC, Canada. Klopfenstein, N. B. *et al.*, (ed.) 1997. USDA Forest Service General Technical Report RM-GTR-297 http://www.fgcouncil/.bc.ca/framdocs.htm.

Glossary of Biodiversity Terms. UNEP-WCMC.

Government of Canada. 1991. "Glossary of selected terms." *The State of Canada's Environment*. Ottawa.

Graudal, L. *et al.*, 1997. Technical Note 48. Danida Forest Seed Centre.

Hagedorn, S. A. An Agricultural and Environmental Biotechnology Annotated Dictionary. Bio Tech Resources. Indiana University. Life Science Dictionary. http://biotech.chem.indiana.

Hagedorn, S. A. An Annotated Dictionary. http://gophisb.biochem.vt.edu/resources/glossary.htm.

Helms, J. (ed.). 1998. The Dictionary of Forestry. Society of American Foresters, USA.

http://www.soils.org.

http://biology.about.com.

http://ecart.iao.florence.it/capstat/grmfr.htm.

http://eire.census.gov.

http://en.wikipedia.org.

http://plantfacts.osu.edu.

http://pmep.cce.cornell.edu.

http://www.1stnclass.com/.

http://www.agronomy.org.

http://www.bio.uu.nl.

http://www.crops.org.

http://www.ec.gc.ca.

http://www.fao.org.

http://www.fertilizer.org.

http://www.fgcouncil.bc.ca/framdocs.htm.

http://www.hortico.com.

http://www.hungrymonster.com.

http://www.lesslawn.com.

http://www.lucidcafe.com.

http://www.oecd.org.

http://www.peopleandplanet.net.

http://www.plantpath.cornell.edu.

http://www.productivityproducts.net.

http://www.stats.gla.ac.uk.

http://www.tealand.com.

http://www.thegardenhelper.com.

http://www.vinesugar.com.

IBPGR (comp.), 1991. Elsevier's Dictionary of Plant Genetic Resources, Rome,Italy.

Iowa State University. Principles of Biotechnology. (Bio-1). http://www.nal.usda.gov/Bic/Education_res/iastate.info/bio1.html.

IPGRI. 1993. Diversity of Development, The Strategy of the International Plant Genetic Resources Institute.

ITTO. Technical Guidelines for the Establishment and Management of Ex situ Conservation Stands of Tropical Timber Species.

IUCN, UNEP, WWF, 1991. Caring for the Earth. A Strategy for Sustainable Living.

IUCN/WWF/UNEP.1981. Conservation Strategy.

Kemp, R.H. 1993. Principles and concepts. FAO Forestry Paper 107.

King, R.C.; Stansfield, W.D.; 1997. A Dictionary of Genetics. Fifth edition. Oxford University Press.

Koski, V.; Skroppa, T.; Paule, L.; Turok, J. 1997. Technical guidelines for genetic conservation of Norway spruce (*Picea abies* (L.) Karst.). EUFORGEN. IPGRI.

Louwaars, 1994.

MacMillan, S. (comp.) 1980. Penguin Dictionary of Biology.

Mátyás, C. (ed.) 1997. IUFRO World Series Vol. 6.

Maynard, C., 1996: Forest Genetics Glossary. SUNY College of Environmental Science and Forestry.

Mayr, E. 1969. In: BioTech Life Sience Dictionary, BioTechResources and Indiana University.

North Dakota State Water Commission. 1988. *Water words: a glossary of water-related terms*. Bismark.

ODA. Silvicultural Manual for the Solomon Islands, Solomon Island Forest Record N°6, ODA Forestry Series No. Wright, J. 1976. Introduction to Forest Genetics. Department of Forestry Michigan State University. Academic Press. (AP) New York.1.

On-Line Biology Book: Glossary.

Parker, Sybil P. (Ed). 1984. *McGraw-Hill Dictionary of Scientific and Technical Terms*. 3rd ed. New York: McGraw-Hill.

Pecs, 1993. SAETTEM.

Schmidt, L. 1997. Tree Improvement Glossary. Technical Note 46. Danida Forest Seed Centre.

Searchable Biotechnology Dictionary. University of Minnesota. http://www.plpa.agri.umn.edu.

Singh, S.S. 1990. *Principles and Practices of Agronomy(Glossary:Water Managemenmt)*, Kalyani Publishers.

Solbrig, O.T. *et al.*, 1992: Biodiversity and Global Change. International Union of Biological Sciences (IUBS).

Somani, L.L. and Tikka, B.S. 1994. *Dictionary of Agriculture.* Agricole Publishing Academy.

SRD Project. 1996. Glossary of Terms relevant to: Sustainable Agriculture, Seed Production and Handling Technology, Genetic Conservation, Agroforestry, Indigenous Knowledge Systems Development and Related Fields and Issues. Ottawa, Canadahttp://gophisb.biochem.vt.edu/resources/glossary.htm.

Temperate Forest Foundation. Glossary. USA. http://forestinfo.org/Glossary.

The Convention on Biological Diversity. Article 2. UNEP 1992.

The Habitat Restoration Group, BioTech Life Science Dictionary, BioTechResources and Indiana University.

UNEP. 1995. Global Biodiversity Assessment. Annex 6, Glossary.

UNEP. Convention on Biological Diversity. Article 2.June 1992.

UNESCO, 1974. World Meteorological Organization. *International Glossary of Hydrology*. Geneva, Switzerland

US Environmental Protection Agency. 1989. *Glossary of Environmental Terms and Acronym List*. Washington, D.C.

USDA Agricultural Research Service. Plant, Microbial, and Insect Genetic Resources, Genomics and Genetic Improvement. Program Component Definitions. http://www.nps.ars.usda.gov.

USDA Forest Service. 1980. A Glossary of Terms for Forest Tree Improvement Workshop. Sacramento, California.

Whittow, John. 1984. *The Penguin Dictionary of Physical Geography*. Markham: Penguin Books.

World Resources Institute (WRI) "What is Biological Biodiversity" http://www.wri.org.

WRI. "What is Biological Diversity". In: "The Diversity of Life" http://www.wri.org.

Wright, J. W. 1976. Introduction to Forest Genetics. Michigan State University, New York.

Zaid, A. *et al.*, 1999. Glossary of Biotechnology and Genetic Engineering. FAO Research and Technology Paper No. 7. ISBN 92-5-104369-8 http://www.fao.org.

Žumer-Linder, M. 1979. Ecological studies, Swedish University of Agricultural Sciences, Uppsala.

Žumer-Linder, M. 1979. Environmental World-List; Swedish University of Agricultural Sciences, Uppsala.